THE SUNDAY TIMES
Guide to the Internet

THE SUNDAY TIMES
Guide to the Internet

Matthew Wall

HarperCollins*Publishers*

Matthew Wall is a freelance journalist and TV research/producer best-known for his weekly *Web Wise* internet column in *The Sunday Times*. He also writes internet features for the paper's Culture section and has written several internet reports for business. In addition, he advises companies on web design and strategy.

HarperCollins Publishers
77-85 Fulham Palace Road
Hammersmith
London W6 8JB

fireand**water**.com
Visit the book lover's website

First published 2000

Reprint 10 9 8 7 6 5 4 3 2 1 0

© Times Newspapers Ltd. 2000

ISBN 0 7230 1073 0

Designer: Sylvie Rabbe
Layout: Beatrice Waller
Editor: Sarah Barlow

Designed, edited and typeset by Book Creation Services Ltd.

Printed and bound in Great Britain by
Omnia Books Ltd, Glasgow G64

Contents

Acknowledgments

I would like to thank my wife, Wendy, for her patience and encouragement during the writing of this book. She kept the coffee flowing and confiscated my games disks – essential interventions for which I am truly grateful. I would also like to thank Christopher Riches at HarperCollins for his flexible interpretation of the word deadline.

Matthew Wall
April 2000

Getting Started

Introduction

Welcome to *The Sunday Times Guide to the Internet*. This book tells you everything you need to know to get started on the internet – in plain English. It filters out all the stuff you don't need to know and concentrates on making the internet work for you, as simply and straightforwardly as possible.

If you've been put off by the internet's geeky reputation, intimidating technology and ludicrous jargon, then this book can help. It reduces the internet to what it really is: a very useful – but sometimes maddeningly frustrating – tool, which can also be great fun.

Our general guide is intended to be a companion to a series of shorter guides focusing on specific topics, such as shopping and money. It isn't a web *site* guide as such – although there are plenty of useful web addresses listed – it's designed to get you up and running, surfing safely and making the most of what the **net** has to offer.

So what's so great about the internet?

- You can write letters and send documents, pictures and sound files across the globe for the price of a local call using electronic mail (**e-mail**).

- You can shop online, book tickets and holidays, and manage your finances from the comfort of your own home, saving time and money in the process.

- You can swap views and ideas with likeminded people from all over the world.

- You can find things out quickly and efficiently using **search engines** and directories.

- You can hear and read up-to-the-second news reports.

- You can **download** music and computer software on to your computer.

- You can play games simultaneously with other people who may even be in a different country.

- It never closes (except when your computer crashes!).

What is it exactly and how does it work?

Who cares? Do you really know how your microwave or mobile phone works? The internet is getting so widespread these days – there are approaching 250 million people online across the world – that it is no longer just a technical novelty beloved of enthusiasts. It is a mass-market means of communication, just like the television. And the amazing thing about the internet is how quickly it has managed to achieve this level of acceptance. It is hard to believe that just five years ago the internet barely registered in the public consciousness.

The Sunday Times website: top stories just a mouse click away (www.sunday-times.co.uk).

Anyway, if you really must know, the internet is basically lots of computers linked up to form networks sending and receiving digital data that can be still or animated images, video or sound. Pretty much anything can be translated into digital data (strings of ones and noughts), then chopped up into small parcels of information called 'packets', and sent across the telephone network. We receive television pictures through an aerial, satellite dish or cable connection, and the same goes for internet data.

The main way people access the internet is through computers linked up to the telephone network, at home and

3

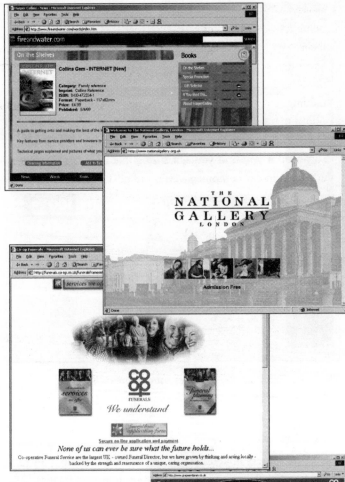

You can organize more and more of your life over the internet: go shopping, check out museums — even arrange your funeral!

at work. But other ways of accessing the internet are rapidly being introduced. Interactive digital television, introduced to the UK in 1999, brought internet access into the living room, and the recent rise of internet-enabled mobile phones and hand-held organizers took it into the street. In fact, we're moving towards an 'internet anywhere' world in which even photo booths, telephone kiosks and bank cash machines are being equipped with internet access facilities. Don't be surprised if you find yourself surfing the net via your voice-activated in-car computer within the next year or two!

One point to make clear is that the **World Wide Web** is only one part of the internet, although it has rapidly become the most important and fastest-growing part.

For a more detailed description of the internet and its history, *see* **Appendix: A Brief History of the Internet**, *page 203.*

The number of ways you can access the internet is increasing dramatically. Dreamcast (www.dreamcast-europe.com) is the first game console to give access to the internet and e-mail.

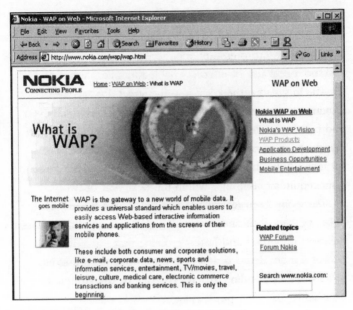

You can even access the internet through a mobile phone using systems such as WAP.

What do I need to get started?

The basic requirements to get the most out of the internet are:

● a computer
● a **modem**
● an internet service provider.

How do I choose the right computer?

Reading computer adverts is like entering a different universe inhabited by people who only talk in acronyms and abbreviations. There are so many variables. The problem with computers is that they are developing so fast. No sooner have we splashed out on the latest model than it has been superseded by a faster one equipped with all the latest gizmos. The whole weight of the industry's marketing muscle is used to pressure us into upgrading even when we may not need to.

The fact is that these days you can buy a perfectly adequate internet-ready desktop computer for around £500, and prices are continuing to fall as competition amongst computer chip manufacturers intensifies. Yes, you can pay a lot more, but often that's for functions and power far beyond your requirements.

There are three main considerations when buying a PC: price, performance and service. The first thing to ask yourself is what you want the PC for. If it is to carry out simple word processing tasks and send e-mails, you're not going to need as powerful a machine as the games enthusiast.

There's also little point in going for a special deal that includes a scanner if you are never likely to use it. So don't

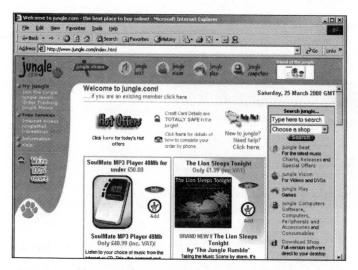

If you have access to the internet through work or a friend, you can purchase your own computer through websites such as this appropriately named jungle.com.

get carried away by marketing hype. You don't need a Ferrari to go to the supermarket.

Mac or PC?

The first choice is between an Apple Macintosh (Mac) computer or what's called an IBM-compatible PC using Microsoft Windows 3.x/95/98 or Microsoft NT as its operating system. Microsoft certainly dominates the market with a 90% market share. But those who use a Mac are a fanatical bunch who swear by its superior design and ease of use.

Generally speaking Macs tend to be preferred by graphic designers and arty people. Certainly, if the look of your computer is important, the Macs beat the boring cream box PC hands down. But if compatibility with everything that's available on the web is more important, you may be better off sticking with the undisputed market leader.

Having said that, the main elements of both type of computer are the same:

The processor

This is the microchip brain of the PC. Its speed is measured in Megahertz (MHz). The larger the number the faster the machine. 450MHz and 500MHz chips are now becoming standard, but as the two main manufacturers, Intel and AMD, compete for market share, some analysts expect 800MHz and 900MHz chips to be on the market before the end of 2000. You really need about 300MHz these days to handle all the latest applications.

> **TIP**
>
> When buying a computer, the first thing to ask yourself is what you want it for. If it is for simple word processing tasks and to send e-mail, you won't need a hugely powerful machine.

Apple Macs: for people who believe design matters

Memory

The more Random Access Memory (RAM) a PC has
the more tasks it can carry out at the same time. If you
don't have enough RAM your PC can start to run slowly
if you have many programs open at the same time. Around
64 Megabytes (MB) is considered enough to handle most
tasks required of a home PC, although highly specified
PCs come with 128MB of RAM as standard. If speed is
important to you, pay a bit extra for more RAM. A lot of PC
retailers allow you to specify how much RAM you want in
your machine.

Hard disk capacity

These days the storage capacity of PCs is measured in
Gigabytes – 1 Gigabyte (GB) = 1,000MB. You need a large
amount of space to store all the software programs you want
to run and any documents, videos, pictures and sound files
you choose to keep. Around 2GB is considered the bare
minimum these days. Most well-equipped PCs will have
around 10GB.

Modem

The good news is that almost all new PCs now come with a
modem built in, so it's just a question of plugging your PC
into the telephone socket to get online. The modem is a
gadget that converts digital data into analogue format for
transmission across the telephone network and vice versa for
incoming data. Modem
speed is measured in
kilobits per second (kbps).
The standard speed is now
56kbps – the higher the
number, the faster you can
send e-mails and

> **TIP**
>
> *A storage capacity of around 2 Gigabytes
> is considered the bare minimum for a PC
> these days.*

download web pages. Just make sure the modem is 'V.90
compliant'. This means that the modem is capable of talking
to all the different makes and models of modem that exist on
the market.

   If you already have a computer that doesn't have an
internal modem you need to buy an external one and plug it
into the back of your computer. There are lots on the market,
some offering fax and answering machine facilities, too. Go
for the fastest and make sure it's V.90 compliant.

   If you want to connect using a laptop or notebook
computer, the newest models also come with modems built
in. Otherwise there is a whole range of credit-card sized

'PCMCIA' modems on the market that simply slot in at the side of the computer. Some just enable you to connect to the internet via the normal telephone network. Others also allow you to connect using your mobile phone, or link up to your company's computer system.

There are other ways of connecting much faster to the internet – using **ISDN** lines, **cable modems** or **ADSL** technology – but these are dealt with in **Faster Surfing**, *page 87.*

Screen size

It can be quite annoying viewing web pages designed for a large monitor on one that is too small. As PC use has become visually richer, screen size and quality has become more important. Basically, the larger the screen the better. Most good PCs will be a minimum of 15 inches measured diagonally from corner to corner, but you can go up to 19 inches or more.

Graphics

If you're interested in playing games on your PC, you'll need a PC with a powerful graphics card capable of handling the very latest, visually rich games that are on the market. Look for the amount of memory the card has – 16MB to 32MB is ideal.

Software

To make their PCs more attractive, manufacturers and retailers will often include software packages – programs that perform specific tasks – with the PC. But just because a PC may come with Windows 98 operating system pre-loaded, don't assume that it will come with all the other software you'll need, too. The operating system software isn't the same as the word processing or spreadsheet software, for example. This is a common misconception.

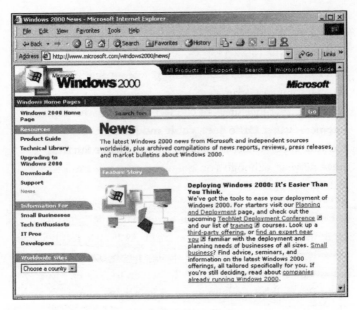

Just when you think you have all the latest software, new versions are released. Windows 2000 is expected to displace 98 as the standard PC operating system.

Buying software separately can add several hundred pounds to the overall cost of the PC, so look very carefully at what's included in the price.

Service

The quality of after-sales service is almost as important as the specification of the PC itself. All machines are fallible. In fact it seems that the more powerful and complicated PCs become, the more likely it is that

WARNING

Look closely at the quality of the telephone support offered by an ISP. Home users are more likely to use PCs after work, so a helpline that shuts at 5pm is not much use. Ideally, you want free, 24-hour lifetime support.

something will go wrong. Often glitches are nothing to do with the computer hardware, but with the software that came with it.

If something does go wrong with your machine, the last thing you want is to have to send it back to the retailer or manufacturer. An 'on-site' warranty, where an engineer will come to your home and mend your PC for you, is best. Most manufacturers' warranties last for just one year, although you can sometimes pay extra to extend the period.

What is an internet service provider?

An internet service provider (**ISP**) is a company that acts as a gateway to the internet. All data coming to and from your computer goes through the ISP first. For example, when people send you e-mail they don't send it direct to your computer. It goes first to your ISP's computers (known as servers) and sits there until you request it. The mail is then sent to your computer.

ISPs can offer several e-mail addresses, free space to create your own web pages, entertainment, shopping and information services, plus technical support. But the best thing about ISPs is that you connect to them at local call rates, even if you're sending an e-mail to California or downloading web pages from Australia. More are offering unmetered calls for a fixed monthly fee.

Most ISPs used to charge a monthly fee of around £10 to £15 for unlimited access to the internet. But this way of charging was blown out of the water by Freeserve, a free ISP service launched in 1998 by Dixons, the electrical retailer. Nearly everyone has followed suit – there are over 200 free ISPs now – about half the total number.

So how do I choose one?

Choosing an ISP depends largely on how you plan to use the internet. Here's a checklist of topics to consider – some may be more important to you than others.

Technical support

This is very important. If you're new to the internet you're likely to need quite a bit of help, especially if there are gremlins in the ISP's software, as there sometimes are. You need to find an ISP with a support line that stays open as long as possible, or at least during the hours when you're most likely to be online. Ideally, look for 24-hour, seven-day-a-week help. But watch out for the cost. Many of these 'free' ISPs charge 50p to £1 a minute and some don't offer any help at all.

Free web space

Many ISPs offer you free space on their servers so that you can design your own web pages. If you fancy doing this, and airing your family photos, doodles, poems and hobbies in public, look for an ISP that offers the most Megabytes of web space. The more sophisticated you want to make your website, the more memory you'll need. Around 10MB to 25MB should be enough. Some ISPs offer unlimited space.

> **TIP**
>
> *When choosing an Internet Service Provider, look in specialist internet magazines for tests that show which are the fastest and most reliable.*

Multiple e-mail addresses

If other members of your family use the PC, look for an ISP that offers several e-mail addresses. That way your private e-mail won't be read by anyone else.

Speed and reliability

The internet is a vast network of computers and as such is only as fast as the weakest link in the chain. Some ISPs are faster and more reliable than others, and if you plan to use the internet extensively, this matters. It can be pretty annoying if you're expecting an important e-mail and you can't access it because your ISP's systems have crashed, or you have to wait ages for web pages to download because your ISP doesn't have enough servers or modems to handle demand.

Specialist internet magazines carry out exhaustive tests on all the leading ISPs, checking them for speed and reliability. A free ISP isn't necessarily less reliable, although, on average, the ISPs that still charge a monthly fee come out best. If reliability is your number one priority, it still may be worth paying for your internet access.

ISP services

Some ISPs offer their members entertainment, shopping and information services (known as 'content' in the jargon) that aren't available to non-members. You have to subscribe and load their software on to your computer. The most popular of these so-called 'proprietary' ISPs are America On-Line (AOL) and CompuServe. If you want to check out their services before committing yourself to a monthly fee, you can usually find their software on free CD-ROMs attached to the front of internet magazines. Bear in mind that free ISPs, such as FreeServe and Virgin Net, also offer their own content on their websites, making it available to members and non-members alike.

Is a particular ISP a member of the Internet Services Providers Association (ISPA)?

It is not essential for your ISP to be a member of the ISPA, but its 110 members do agree to stick to a code of conduct

You can find a full list of ISPA members on its web site (www.ispa.org.uk).

and there is a standardized complaints procedure if you're unhappy with your ISP's level of service.

What if I'm a Mac user?

The brutal fact is that computer life is dominated by Microsoft in almost every area. Until the breakthrough introduction of the iMac it wasn't easy to get online with a Mac at all. Even though there has been a vast improvement, many ISPs are still not geared up to give technical support to Mac users. Many websites are not designed to be used by Macs. Before you sign up with an ISP check that they are fully Mac-friendly. Once online, there are plenty of Mac-related websites to help you navigate the net in a way most suitable for your machine. Try these for starters:

www.apple.com

www.macobserver.com

www.macuser.co.uk

www.everymac.com

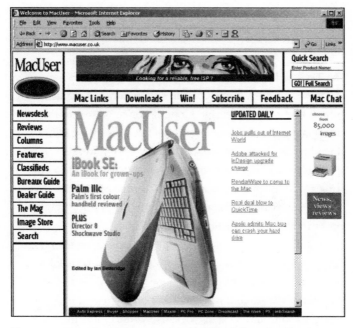

There are plenty of web sites around dedicated to helping the Mac user.

Can I have more than one ISP?

There's nothing to stop you having links to several ISPs on your computer. In fact it can be a good idea to have a back-up connection if one breaks down for some reason. It can also be practical to have one e-mail address for business and one for personal use. If you've set up your e-mail address with one ISP you can even access your mail using another ISP. The only thing you can't do is *send* e-mail from a competing ISP.

More often than not new computers will come with software for one or two ISPs already loaded. You don't have to use them if you don't want to, but there's no reason to get rid of them either. They can be useful for back-up.

Do I need to install a new telephone line to get online?

No, but it is a good idea if you can afford it. As the internet uses the telephone system anyone trying to ring you will just get an engaged tone if you're online. Surfing the internet can become addictive – you can find yourself spending hours online without realising it – and this can be intensely annoying for people trying to get through to you. This also applies if someone wants to send you a fax. If you have a large family, hogging the telephone line for hours on end can also cause friction.

Another good reason for installing a dedicated telephone line just for internet use is that you can talk to your ISP's technical support staff on the phone *and* connect to the internet at the same time. This can be very useful if you're trying to sort out problems, because you can test their suggested remedies without having to hang up.

Hanging up can be bad for your health. Helplines often use an automated telephone queueing system that directs your call to whoever's available. So if you've spent a long time explaining your internet problem to one technician, it can be extremely frustrating having to explain it all again to another technician when you ring back. An extra line also helps you keep a tab on your internet call costs, which can mount up alarmingly, especially if you surf extensively during the day, when call costs are most expensive.

TIP

If you have a second telephone line installed for internet use, make sure you tell the telephone company that this is what it is for, so they can supply you with the right sort of connection.

Not all lines are the same

If you do go ahead and get a second line installed purely for
internet use, tell the telephone company that this is what you
want it for. Otherwise the company might just fit a device that
lets two telephone lines share the same connection back to
the exchange. This is called Digital Access Carrier Service
(DACS). It can have the effect of halving your modem speed,
leading to slower download times and a more frustrating web
experience.

Making one line go further

If installing a second line is too much bother or expense,
there are software packages available now that can monitor
your telephone connection while you're online. For example,
Online Call Manager (**www.ocm.uk.com**), from Redstone
Telecom and Witchity Capital Corporation, tells you if
someone is trying to ring you while you're online. The
software gives you the option to handle such calls in a

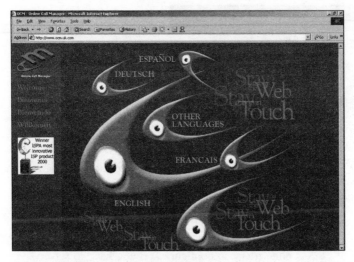

*Software packages such as Online Call Manager (OCM, www.ocm.uk.com)
can monitor your telephone connection while you're online.*

number of ways. For example, you can transfer them to your mobile phone or ask the caller to leave a voicemail. Of course, you have to pay for such services, although the software is free to download. So it's a question of weighing up how much you'd save by not installing a second line and paying the monthly line rental on it against the running costs of such a call manager.

There are other software programs that can help you make the most out of your single telephone line. Many homes these days have more than one PC, but only one may have a modem installed. You could install a second line and buy a second modem, but that would be expensive and potentially a waste of time given that high-speed connections should soon be the norm (more on **Faster Surfing**, *page 87*).

Now there are software solutions to help you give net access to both PCs without having to install a second line and buy a second modem. Microsoft's Internet Connection Sharing (ICS) software enables your wired-up PC to act as a gateway to the net for your other PC. Two members of your family could be online at the same time, one sending e-mail from one PC, say, the other surfing the net on the other. As they'll be sharing the same connection it does mean they may experience slower connection and download speeds, especially if both are downloading files and programs at the same time. For normal surfing though, the difference should be barely noticeable.

ICS is quite tricky to configure and you have to set up both computers so that they are 'networked'. This involves making sure there is a network interface card in each PC and

> **TIP**
>
> *You may find it cheaper to use a call manager software program than to install a second telephone line exclusively for internet use.*

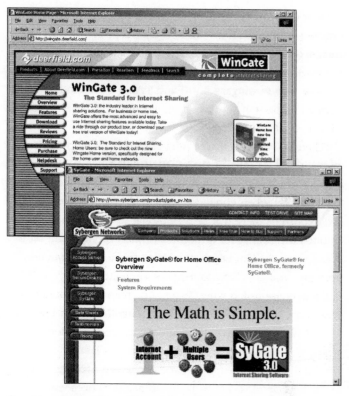

Two alternatives to ICS are Wingate (www.wingate.com) and Sygate (www.sygate. com).

that they can handle TCP/IP (*see* **Glossary**). Another drawback is that ICS is only available in the full retail and upgrade versions of the Windows 98 Second Edition and Windows 2000. To install ICS you go to 'Control Panel', 'Add/Remove Programs', 'Windows Setup', then scroll down to 'Internet Tools', highlight it, then click on the 'Details' tab. Check the box next to 'Internet Connection Sharing', click 'OK', then 'Apply'. After that, have your Windows 98 SE installation disk at the ready and follow the on-screen prompts which take you through the installation process.

If you encounter problems, there are plenty of ICS-related articles on **Microsoft's** website (http://support. microsoft.com/search). There are also alternatives to ICS on the market, such as **WinProxy** (www.winproxy.com), Wingate (www.wingate.com) and **Sygate** (www.sygate. com).

How do I actually get connected?

Normally when you're installing the software offered by your ISP, there will be a step-by-step on-screen guide through the connection procedure. The most important bit is telling your modem which number to dial to get through to your ISP. You also have to tell your computer what your ISP's computers are called when setting up an e-mail account (for more on e-mail *see* **Electronic Mail**, *page 103*).

Usually your ISP will have an icon on your desktop which you click on to trigger the automatic dial-up process. If your modem is on and connected properly to the telephone socket, you should hear a faint 'peep peep peep' as the number is dialled. You then hear a combination of bizarre whooshing sounds and beeps, just like the burble a fax makes, as the computers try to talk to one another.

Depending on the ISP, you then go through an online security check of your user name and password – you usually set these up when loading the software. Such security can be quite useful to prevent people logging on to your ISP in your name and sending unflattering e-mails to your boss, for example, or, perhaps worse, reading unflattering mail sent to you.

Depending on how your ISP has configured its software, the web browser should automatically open at this point. But you don't have to do it this way. You can click on the browser icon on your desktop first, open the browser window, and then decide which of your ISPs you want to connect to (remember, there are loads of free ISPs available now, so you can run several ISPs alongside each other at no extra cost).

If you have a 'connect automatically' box ticked, the dialler will just go ahead and dial through to your default ISP. If not, you can select your chosen ISP at this point before clicking on the 'connect' button.

What if I can't connect?

Unfortunately, there are number of things that can go wrong when trying to connect. My advice is to phone your ISP first if you encounter problems, even if you have to pay handsomely for the technical support. This is better than fiddling with settings yourself if you don't really know what you're doing. Speaking personally, I have found the 'Help' sections that come with software packages next to useless.

Your ISP should be able to take you through all the possible causes of the problem step by step. Some common reasons for a failure to connect are:

● **No modem detected –** if you have an external modem it may not be switched on or it may be plugged into the wrong socket at the back of your computer. It may also mean that the software accompanying the modem wasn't installed correctly. You may have to re-install it.

T I P

A number of things can stop your modem from connecting. Try running through this checklist, but if you still can't get online, your ISP should be able to advise you.

❷ No dial tone detected – it could be that the telephone
cable isn't inserted into the wall socket properly, or that
the socket itself is faulty. If you're using a 'splitter'
connector so that you can share the socket with a
telephone, the connector itself may be faulty. Check that
the line is OK by inserting a telephone jack into the socket
and listening for a dial tone. If you still don't hear
anything, there's a fault on the line.

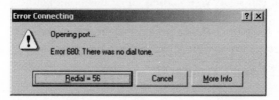

❸ Number engaged – at busy times of the day it is
sometimes difficult to get through to your ISP. Good
ISPs will have several numbers for you to use should
you fail to get through on the main one. The simplest
thing to do is leave it for a couple of minutes and
then try again.

❹ Password not recognized – when installing your ISP
software and going through the set-up and registration
process you normally have to enter a user name and
password. This is often assigned to you and many free ISPs
don't bother with this. If you've entered the password
incorrectly, and then told your computer to remember the

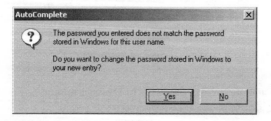

incorrect password, you'll never get through! Try retyping the password in the 'Dial-up connection' box, taking care to get it exactly right. Sometimes it's not your fault at all and there's a glitch in your ISP's security systems. Again, it's as well to check with them.

⑤ **Protocol problems** – sometimes your modem will have difficulty 'talking' to your ISP's modem. This could be to do with your **TCP/IP** settings – apologies for the jargon, but it's unavoidable sometimes. TCP/IP (Transport Control Protocol/Internet Protocol) is the standard way computers talk to each other on the internet. Your computer has a unique **IP address** – usually four sets of digits separated by dots – as does every other computer on the net. This helps all the information flying across the global network to arrive at the right place. Computers called Domain Name System (DNS) servers attribute names to the numbers to make the net a friendlier place. You may have to enter your ISP's exact IP address – ask your ISP

> **TIP**
>
> *Make a note of your IP settings for each ISP so that you can re-enter them if you need to.*

for help on this one. A word of warning: sometimes when you load another ISP's software on to your system it can interfere with the IP settings you've entered for your other ISP(s). If this happens, you'll have to re-enter them. It's a good idea to make a note of them somewhere for future reference. It will save a potentially expensive call to a helpline.

⑥ **You connect, but it keeps disconnecting for no reason** – this often happens if you have 'call waiting' on your line. The beep that you hear when you're on the phone normally interferes with the net connection and your modem drops the call. Ask your telephone company how to activate and deactivate call waiting. Another reason may be that your web browser settings are configured to drop the connection if there's a period of inactivity. The website – an online bank, for example – may do this too for your own security.

Chapter 2

Web Browsers

What is a web browser and why do I need one?

A web **browser** is a software program that helps you surf the net. Web pages are downloaded into the browser's window. You can type in web addresses and go to different sites, go backwards or forwards through web pages, and store your favourite website addresses in directories that you can organize in your own way. They can provide access to electronic mail, news and chat services. And the latest versions of web browsers will also help you screen websites and check them for authenticity and security.

When an internet service provider sends you a CD-ROM containing its software, most of it is actually the browser software. Although there are several web browsers to choose from, by far the most popular are:

- **Microsoft Internet Explorer** (IE) – www.microsoft.com
- **Netscape Navigator** (its enhanced version is called **Netscape Communicator**) – www.netscape.com

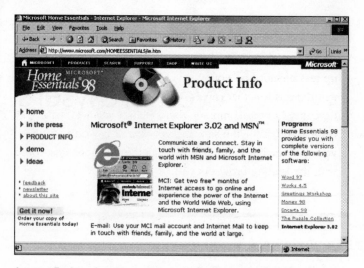

Internet Explorer from Microsoft is currently the most widely used browser.

IE is now the dominant browser, with over 50% of the market, having seen off its rival Netscape quite convincingly. Navigator is generally preferred by Mac users, although IE is usually pre-installed on new Macs these days. Other simpler browsers include:

- **Opera** (not available for Mac users) – **www.opera.com**
- **Lynx** (a text-only browser) – **www.lynx.browser.org**
- **1X** (a new stripped-down browser still capable of handling animated graphics on the net) – **www.scitrav.com**

TIP

If you want to find information on the web and are not interested in graphics, you may find that a text-only browser suits you best. It will save you time, as it is much quicker to download words than images.

You may wonder, what is the use of a browser that can't handle pictures, given the net's penchant for graphics? Well, graphics take up a lot of memory and downloading them takes time. If it's information you're after, and you want it fast, a web browser that ignores graphics will be a lot quicker.

Also, the alternative browsers tend to be simpler, taking up less memory on your hard drive. In theory, this means that there is less that can go wrong. It also means that if you have a rather old computer, there is more chance that it will be able to handle the simpler browsers. It is worth asking yourself whether you really need all the functions that come with the leading browsers.

Can I have more than one browser?

Yes. There's nothing to stop you having a number of browsers on your system – you don't have to stick with the browser your ISP has given you. CD-ROMS accompanying computer and internet magazines often contain the latest versions of the various browsers available. You can load them on to your computer and try them out to see which you prefer. It is far quicker to load software this way – downloading browsers from the net can take hours, especially with a slow modem.

All software programs can develop glitches from time to time, so it's just as well to have a back-up. What you do tend to find is that each browser will try to make itself the 'default', or main, browser when you load it. When you try to connect to the net the 'default' browser program will automatically launch. But you can switch between browsers relatively easily. These days it is also possible for different browsers to read your collection of favourite websites (called 'Bookmarks' in Netscape Navigator and 'Favorites' in Microsoft Internet Explorer).

In my view there's not much difference between all the various browsers, so it's really just a question of personal preference.

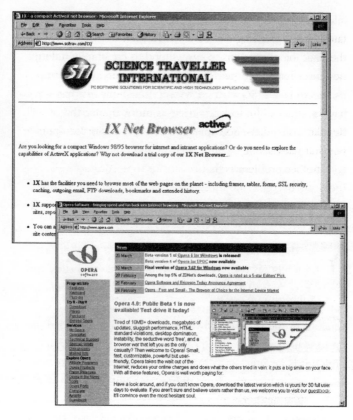

Popular alternatives to IE and Netscape include IX (top, www.scitrav.com) and Opera (bottom, www.opera.com).

Keeping your browser up to date

The problem with software companies is that they keep bringing out newer versions of their browsers (usually denoted by an ascending numerical series – 4.1, 4.2 etc). And as website design becomes more sophisticated, incorporating animation and sound, for example, we need the latest versions of browsers to read these pages properly.

Many people encountered problems trying to use the latest online banking and stockbroking web services because they had older versions of browsers that couldn't handle the new user name and password security features introduced by the websites. So when you've chosen your browser(s), it is very important to keep a look-out for upgrades. Regular visits to the software company's website for upgrade news are advised.

Another problem is that the more sophisticated web browsers become, the more sophisticated computers have to be to run them. Some of the newest versions of browsers are not compatible with older operating systems. So you have to find the browser version that is most compatible with your operating system. Your ISP should be able to give you advice. Yes, it is fiddly and annoying and all part of the general conspiracy to keep us buying new computers and new software. But the net is developing at

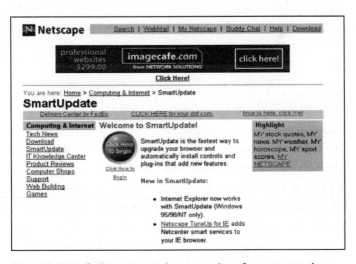

Keep a look-out for browser upgrade news on the software company's website.

such a pace that to make the most of what it can offer does require the latest hardware and software. That's just a fact of life online. This book assumes that you have the latest browser versions anyway.

Getting to know your browser

Y ou should get to know your browser intimately because it's the main tool you use to surf the next. It's there to make surfing the web easier and faster for you, so the more effort you put into discovering all its functions, the more rewarding your experience will be.

The web address bar | Address 🔊 http://www.sunday-times.co.uk/

This is the white bar or box usually at the top of the web browser. It's where you type in web addresses that take you to specific sites. In Internet Explorer the word 'Address:' is next to the box; in Navigator, it says: 'Go to'. Once it's found the site it says 'Location' or 'Netsite'. Other browsers may say 'URL' next to it. You just type the address accurately in the box, making surely you're at the start of the box, press return, and away you go. For further information on web addresses, *see page 36.*

The 'Forward' and 'Back' arrows ⇐ Back ▾ ⇒ ▾

The clever thing about browsers is that they store the pages you visit while you're online. So if you flit from website to website, it is possible to go back to one you've left, simply by clicking on the 'back' arrow button in the browser's task or menu bar. Similarly with the 'forward' arrow button. It's exactly like flicking backwards and forwards through the pages of a book. There's no need to retype web addresses each time.

Opening new browser windows

An even quicker way to flick between different websites is to open a new window in your browser. In the latest versions of both leading browsers you do this by clicking on the 'File' tab, selecting 'New' and then 'Window' or 'Navigator Window'. The same web page you're on will then load into this new window. But if you type a new address into the address bar, or select a new bookmark, the new website will load into this window. The old website is still loaded in the other browser window.

You can then switch quickly between the websites by clicking on tabs at the bottom of the screen. And if you want to be really clever, you can minimise the browser screens and arrange them so that two or more browser windows are side by side, allowing you to compare and contrast websites, or absorb information from several sources at once.

> **TIP**
>
> *It is quite common for people to read about a website in the paper, say, write down its address, then enter that address into a search directory, such as Yahoo! or Excite. You don't have to do this! The address will take you directly to the site.*

The 'Stop' button

This humble button is actually very important because it stops you wasting time. When the web is busy it can almost grind to a halt, especially when downloading graphics, which take up a lot of memory. If you get fed up of waiting, just click the stop button and move on to something else. If you don't, your browser and modem will continue labouring away trying to download the pages in the background while you're trying to look at another website, slowing up the process even further.

The 'Refresh' or 'Reload' button 🔲

Sometimes web pages don't always load properly first time –
some of the data gets lost on its travels leaving an incomplete
picture or piece of text. If you click on the 'Refresh' or
'Reload' button your browser will simply try to download the
page again, hopefully with more luck. Another use for this
button is to keep web pages up to date. Some websites, such
as financial data providers, provide information that is
constantly updated – share prices, for example. If you are
online for a while, you should refresh the page every so
often – if it doesn't do so automatically – and you'll then
receive the very latest information. Some websites will make
your browser do this automatically.

Other useful functions

As Microsoft and Netscape compete with each other, they
bring out newer versions of their browsers, laden with yet
more features. Most of them are based on the bookmark
principle, giving you short-cuts to web services, such as radio
stations, shopping sites, search directories. Really, there's no
reason why you can't find these services yourself without
having dedicated buttons built into the browser software.

So here's a round-up of the browser features you *will* find
useful:

- **Home** 🏠 clicking on this button takes you back
  to the first page that loads when you first log on to the
  net. Whenever you load an ISP's software they usually
  make their own website your default home page.
  But you can choose any website you like. Or you can
  choose to start with a blank page. In Internet Explorer
  go to 'Tools', 'Internet Options', then type in the web
  address of the site you would like as your home page.
  In Navigator you click on 'Edit', 'Preferences' and follow
  the same procedure.

- **The Print button** 🖳 once you've found an article or picture you like you can print it off there and then.

- **E-mail button** 📧 this enables you to launch your e-mail software from the browser (for more on e-mail, *see* **Electronic Mail**, *page 103*).

- **History** 🕒History you can see all the websites you have visited recently. This is useful if you came across a good site but forgot to bookmark it. You can specify how long you want your History folder to keep records for. If you specify '0' days, it will only show you the websites you've visited today.

- **Security** – the latest version browsers have a whole host of security features to help you filter out websites you deem unsuitable and check whether the sites you visit offer secure, encrypted transactions. *See* **Safe Surfing**, *page 159*, for more on security issues.

- **Save** – as well as being able to bookmark pages so you can return to them when you're online, you can also save specific web pages to read offline. In the latest browser versions you get a choice as to the way you want to save the page. For example, you can tell your browser to save just the text, not the pictures, or the whole caboodle, animated graphics and all.

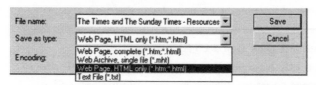

- Another way to **capture text** on a web page is to highlight it by clicking and dragging over the desired paragraphs using your mouse. Once you've highlighted the text, right click on it, choose 'Copy', open a new word processing document, then choose 'Paste' to insert the text into the new document. You can also save specific images by pointing at them, right clicking, then choosing 'Save as'.

● **Send** – if you come across a website that you'd like to tell someone else about, you can send the link or the entire page to them by choosing 'Send' from the 'File' menus of both types of browser. You then send an e-mail to the recipient, so you'll need to know their e-mail address (*see* **Navigating the Net**, *page 45*, for more on e-mail).

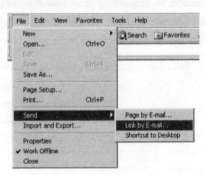

Web addresses explained

Web addresses need some explanation. A web address is a combination of words and symbols that directs you to a particular website, or particular page on a website. The **www** bit of a web address is probably the most widely recognized symbol of the net. Not many companies advertise without including their web address these days.

You may have heard people talk about **URLs – Uniform Resource Locators**. It's just net-nerd-talk for web address. If you're new to it all, the motley collection of dots, dashes, hyphens and abbreviations that makes up a web address can be off-putting. So let's take an example and break it down into its constituent parts.

The Sunday Times newspaper's web address is:
http://www.sunday-times.co.uk

Each colon and forward slash is essential – just a missing dot and your browser will tell you it couldn't find the website. Some Web addresses may also be case-sensitive, so be careful not to put in any capitals where they don't belong.

The **http://** bit tells you that the site is on the World Wide Web. You don't need to know this, but **http** stands for Hypertext Transfer Protocol. It's just another of these web standards agreed by programmers and designers and is not worth worrying about. All website addresses begin with **http://**. So when you see Web addresses written down like this: **www.randomname.com**, it is assumed that you'll type in the **http://** bit first.

Thankfully these days the latest versions of browsers will automatically put in the **http://** for you if you forget. They will also guess which web address you're beginning to type and try to complete it for you. If it's a particularly well-known site you can even type just the main part of the name and the browser will fill in the rest of the address for you.

Ironically, although the **www** bit is the most widely recognised symbol of the internet, not all web addresses include it. Normally the name of the company or organisation comes after the **www** or **http://** bit, followed by an abbreviation which tells you broadly what kind of organisation it is. The most common abbreviation is **.com**, which is short for commercial and is used predominantly by US companies.

In the UK, companies commonly have **.co.uk** after their names; German websites may have **.co.de**, but there's nothing stopping UK or German companies having **.com** after their names either. It can get very confusing. So you shouldn't assume you know where a company or organisation is based just by its web address. The whole point about the net for many is that it is global and crosses national boundaries.

Educational institutions often include the abbreviation **.ac** for academic, and non-commercial organisations may have **.org** in their address. Government departments will usually have **.gov** in there somewhere. Another common abbreviation is **.net**.

Address 🖉 http://www.sunday-times.co.uk

Address 🖉 http://www.nationalgallery.org.uk

Address 🖉 http://www.lib.berkeley.edu

Address 🖉 http://www.louvre.fr

Some typical web addresses.

These names and the accompanying abbreviations are called **domain names** and the system for organizing and classifying them all is called the **Domain Name System (DNS)**. The computers that translate all these names back into **Internet Protocol addresses** – that unique set of identifying numbers mentioned in Chapter 1 – are called domain name servers.

It's a thorny issue as to who should control the available names and abbreviations. One thing is certain, if the number of websites keeps increasing at the same rate we will need more of them.

You can register any domain name you think of with a number of agencies. Some bright sparks registered hundreds of names that companies and organizations were likely to want. Some domain names have changed hands for millions of dollars, such is the perceived importance of having your company's brand name in your web address. But you have to be careful. Just registering names that are very similar to big brand names in the hope of making an easy buck could land you in hot water. It could be interpreted that you are passing yourself off as the company itself for possibly fraudulent reasons.

Once you've typed in the website address accurately and pressed return, you should go straight through to the website's **home page**. This is the opening page of the website that usually tells you about the company or service and what it offers. Each website can have hundreds of pages, and each of those separate pages can have a different address.

If you go through to *The Sunday Times* home page, it looks like this:

A typical web page will have a menu of items on the left taking you to different sections of the website.

Non 'http' addresses

The web is only one part of the internet. Two other significant parts are **FTP (File Transfer Protocol) sites**, mostly dedicated to distributing software across the net, and **Usenet**, the collective name for news discussion groups. FTP web addresses start **ftp://** and Usenet addresses begin with an abbreviation such as **alt.** – for more on this, *see* **Making the Most of the Web**, *page 129*.

Bookmarking websites and web pages

Each page on a website has a unique address that usually gets longer and longer the further you explore a website. If you wanted to go back to that page at a later date, one way would be to write down the exact address of that page, with all its slashes, abbreviations and so on. This isn't practical, to say the least. The likelihood of a mistake is very high and it also wastes time.

If the website has been designed using 'Frames' – a way of organizing the page so that there is a fixed frame of information enclosing pages that can change within the frame – the web address will not change even if you click through to different parts of the website. This could make finding a particular page again even more tricky.

Luckily, there is an easy way to store website addresses. Bookmarking is the best thing your web browser can do. Instead of having to remember ludicrously long web addresses you can just tell your browser to remember the page you're on. With Netscape Navigator, you click on the 'Bookmarks' pull-down menu and click the 'Add Bookmark' option. With Microsoft's Internet Explorer you click on the 'Favorites' pull-down menu and then click on 'Add To Favorites'. You can also click on the right mouse button while pointing at the page and you'll be offered the same options. Or you can just hold down the 'Control' button and press 'D' on your keyboard and it will bookmark the page in both browsers.

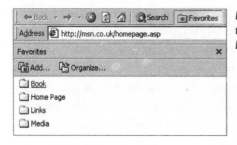

Bookmarking pages is the best thing your browser can do.

Bear in mind that when you do this you will bookmark the precise page you're on. So if you like the site as a whole, and not just the specific page you happen to be looking at, go back to the **home page** first before bookmarking.

Working offline

When bookmarking websites you can also tell your browser to store all the pages on your computer's hard drive so that you can view them again without having to go online. This is a useful feature that can help cut down your telephone bill. Of course, it is best suited to sites whose information isn't likely to change very often, otherwise you may end up reading stuff that is out of date. To prevent this happening you can also tell your browser to check for updated versions of the saved pages next time you go online. You can then save the new pages for offline viewing later.

You can also ask a website called a **notifier** (such as **www.notifier.com**) to tell you when a particular web page you've bookmarked is updated. You just tell the notifier your e-mail address (*see* **Electronic Mail**, *page 103*) and the URL of the page you want monitored.

Organizing your bookmarks

Bookmarking is essential for navigating the net, especially as it becomes ever more complicated and crowded. The latest versions of web browsers also allow you to organize your bookmarks in any way you want. For example, if you're particularly keen on wine websites, you could create a folder called 'Wine' and every time you visit a wine site you like, bookmark it to this folder. Obviously, the same goes for any subject. In Internet Explorer you choose 'Organize Favorites' from the 'Favorites' menu. In Netscape you can choose 'Communicator', 'Bookmarks', 'Edit Bookmarks' or, more directly, click on the dedicated 'Bookmarks' tab and then choose 'Edit Bookmarks'.

When you tell your browser to bookmark a page, it will often give the file a very long title – whatever the site uses to introduce itself. For example, if you bookmarked a fictitious wine site called DrinkYourselfStupid.co.uk, the bookmark title

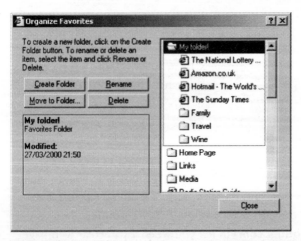

Organizing your favourite bookmarks in Internet Explorer.

might say: 'Welcome to Drink Yourself Stupid, the UK's premier wine website for bibulous buffoons'. You don't need all that, so you can shorten it to Drink Yourself Stupid or DYS if you want.

If you want to return to a favourite site, all you do is click on the 'Bookmarks' or 'Favorites' tab located in your browser's menu bar. A pull-down menu will list all your bookmarks. Just click on the one you want and you'll be whisked straight through to the website.

Understanding web pages

Web pages are generally written in a computing language called **HTML – HyperText Mark-up Language**. That's all you really need to know! They usually look pretty much like standard magazine pages, using pictures, text and colour. But they can be much more dynamic than conventional pages because they can also accommodate moving pictures and sound.

They use **hyperlinks** which are bits of text or images that take you somewhere else once you've clicked on them, or

```
<HEAD><TITLE>The Sunday Times </TITLE>
<!—Main frame set for Sunday Times Front page—>
<!—Day_root /sti/2000/03/26/—>
<!—Issue 9161—>
</HEAD>

<frameset BORDER="0" cols="*,640,*">
        <frame  SRC="/standing/shared/bg.n.html" MARGIN-
WIDTH="0" MARGINHEIGHT="0" FRAMEBORDER="NO">

<FRAMESET COLS="95,*" BORDER="0">
<FRAME NAME="left" SRC="/news/pages/sti/2000/03/26/lefttoday.n.html"
MARGINWIDTH="0" MARGINHEIGHT="0" FRAMEBORDER="NO"
SCROLLING="auto" NORESIZE>
<FRAME NAME="frame2"
SRC="/news/pages/sti/2000/03/26/frame2fpsti.n.html"  MARGIN-
WIDTH="0" MARGINHEIGHT="0" FRAMEBORDER="NO">
```

A sample of HTML text, the hidden coding that underlies most web pages.

launch a snippet of audio or video. For example, if you're on a web page and you see a link with a little speaker icon next to it, you know that if you click on the link you'll hear a snippet of sound – whatever that may be. You need the right software to make the most of these **multimedia** features, and this is dealt with in the next section.

Most text hyperlinks – now just called **links** – are easy to spot because the text is usually a different colour and underlined. But web design has moved on apace. These days links can change colour once your mouse pointer lands on them. You might see a drop down menu appear, for example, containing other links to other parts of the website. When you're pointing at a link you will also see more information about it given in a bar at the bottom of your browser. This might be explanatory text or the web address of the site you'd go to if you clicked on the link.

Another way of knowing that you're on a link is that your mouse pointer will change shape to a pointing hand icon. This is especially useful when pointing at images that are links. When the pointer icon changes to a hand you know that if you click on the link you'll be whisked off to another web page. If the icon doesn't change, the image isn't a link.

Chapter 3

Navigating the Net

Introduction – Big and getting bigger

So now you're up and running and connected (we hope) to the net. That's the easy part. Finding your way around it and pinpointing the information or service you want is much harder. At the last count, there were over eight million websites and over one billion pages on the World Wide Web, with hundreds more being launched every day. As the global internet population is expected to reach 500 million over the next few years, the number can only increase as more services go online to cater for the increase in demand.

Once you've got to grips with your browser you'll want to find out what's out there. After all, it's no use knowing how to type in web addresses if you don't know the address in the first place! If you're after something specific, navigating your way through this cyber equivalent of the Tower of Babel can be tricky, if not exasperating.

In this chapter we'll show you that acquainting yourself with just a few basic rules can help you get to where you want to go a lot faster. But bear in mind that part of the fun

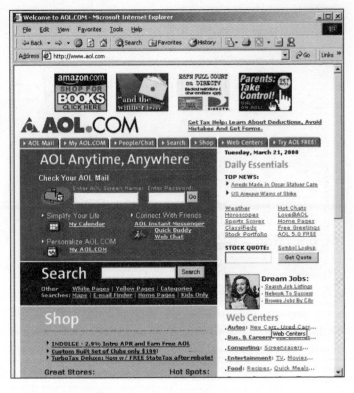

of the net is surfing pretty aimlessly just to see what's out there. You never know what you may find.

Keeping it simple

The net can be as complicated or as simple as you like. Plenty of net users exist quite happily just making the most of the sites listed on their ISP's home page. A good ISP tries to make sure that there is enough to inform and entertain its users, if only to prevent them flitting off to other sites to the annoyance of its advertisers.

Proprietary ISPs, America On-Line (AOL) and CompuServe, try even harder to keep their members within an exclusive environment. The belief is that members will be prepared to carry on paying monthly subscription fees

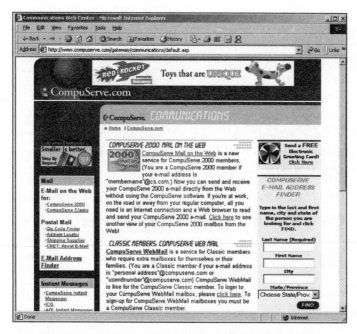

AOL (opposite, www.aol.com) and CompuServe (above, www.compuserve.com) are two of the largest proprietary ISPs.

if they feel they are receiving services not available elsewhere on the web.

Newspapers are increasing their internet coverage by the day, with website reviews and recommendations, and internet and PC magazines also contain tons of website addresses for you to try out. The secret of successful surfing is remembering to bookmark good websites when you visit them.

Before we even get on to using search engines and directories, it's worth pointing out that you may be able to find what you want just by using your browser and guessing. For example, if you're looking for a well-known company's website, it's a good bet that its name is included in the web address. So typing **www.companyname.co.uk** into the address box and hitting 'return' may well get you to where you want to go.

Even if you haven't got the address exactly right, it's likely that the company will have registered all the web addresses that look similar. Your incorrect attempt could still be forwarded to the correct website. Even if you're not this lucky, a little fiddling around with alternative endings, such as **.com** or **.net**, coupled with the odd hyphen between words, will often do the trick.

Using search engines and directories

Having said all this, the first port of call for most people is a **search engine** or **directory**. In theory, these are websites that help you find anything that's on the net just by keying in a few search words in a search box. The truth is less satisfactory. There is simply so much available on the net these days – from the erudite to the downright dangerous – that finding precisely what you want has become an often frustrating and time-consuming process.

Even though search facilities have improved a great deal, I have to confess that I'm not a great fan of them in general. I'd much rather someone tell me the web address of a specific recommended site than root around on a search engine that may or may not find what I'm looking for. **Yahoo!** (www.yahoo.com) may have started out life as a search engine, but it quickly realised that there was far more mileage in becoming an all-singing all-dancing **portal** site. It now has a bigger stock market value than Boeing, the aircraft manufacturer, but

search engine – a program that interrogates the net for files or web pages containing or relating to words or phrases you enter into the search box.

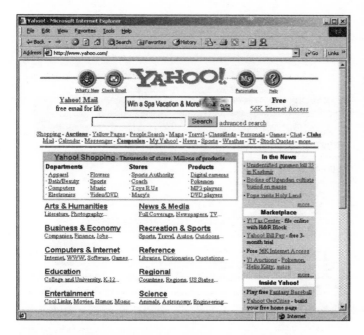

One of the earliest search engines, Yahoo! (www.yahoo.com), has now evolved into a giant portal site.

some critics say it has virtually given up on the idea of trying hard to index the web.

A portal is a general-purpose website that usually contains its own entertainment and information services as well as links to many others. They tend to aggregate services so that their sites will become a destination in their own right, so they are sometimes called **aggregator** sites, too. The search engine or directory often seems like an afterthought.

portal – *a general-purpose website that may contain its own entertainment and information services as well as links to many others. It usually incorporates a search engine or directory.*

One common misconception amongst people new to the net is that *everything* is online, as if the net

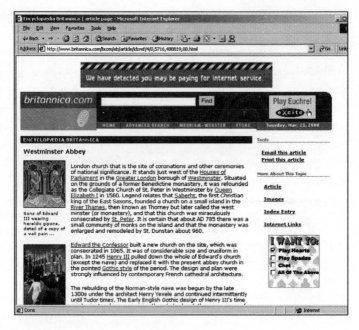

The Encyclopaedia Britannica *(www.eb.com) site.*

has magically put the world's knowledge base online. This is far from the case. Stuff is on the net only if someone has been bothered to put it there. It still has a long way to go before it rivals the resources of a national library, for example. But it is getting there. Even the *Encylopaedia Britannica* has succumbed to the pressure and disgorged its contents on to the web.

Comprehensiveness versus ease-of-use

For the purists out there, a search *engine* is different to a search *directory.* An engine involves a program called a spider or robot scouring all websites it can get access to, recording all the information, and depositing it in huge databases. These spiders can search as many as a thousand pages per second. When you type in a word or phrase the search engine will throw up all documents containing those words or phrases.

This is a very powerful tool, given the amount of data available on the net, especially if you are carrying out academic research and thoroughness is essential. The problem is that if you are not specific enough in your choice of words you can still end up with millions of irrelevant documents that are little help to man or beast. Sifting through them can take hours, many of them wasted. This is not made any easier by the fact that there are now an estimated 800 million web pages on the net. In 1995 there were just 100,000 websites.

Search directories take a more selective approach, with people deciding what is and isn't worth categorizing for their users. Rather than carrying out blanket searches you can narrow down the field by pre-selecting a category, such as Sport or Shopping. This makes it much more likely that you'll find a site that is relevant and interesting. But this kind of indexation does mean that you cannot be sure you've found everything that's out there on the subject you're interested in. You'll only be shown what the search directory company has decided to record and categorize.

> **TIP**
>
> *No search engine lists all websites, so if you look for a site and don't find it, it doesn't mean it isn't out there. Try using a different search engine!*

And there's the rub. Search directories have realized that indexing and categorizing everything on the net is well nigh impossible. It comes as a shock to many to learn that even the biggest search directories cover just 20% of all websites on average. Now this isn't a problem if they correctly guess what surfers are most likely to be interested in and what sites aren't worth including at all. And let's face it, there is a lot of

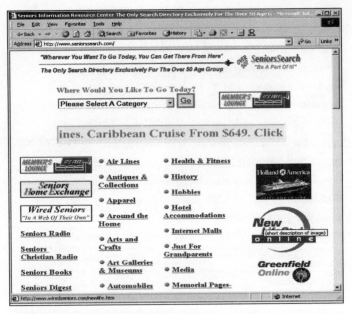

Seniors Search (www.seniorssearch.com) is a search directory dedicated to senior citizens.

dross out there. But comprehensiveness is obviously sacrificed on the altar of speed and convenience.

These days the distinction between engines and directories is becoming blurred, and you'll often hear the terms used interchangeably. In this book, for instance! Despite the rise of UK-specific directories, there is still a bias towards US-based websites, which can be frustrating. And many websites complain that, even though they register with the leading directories, it is a lottery as to whether their company will come up in a relevant search.

The trend is towards more and more specialist directories – a natural and welcome response to the burgeoning size of the web. For example, **Seniors Search** (**www.seniorssearch.com**) is a directory dedicated to collating sites that may be of interest to the over-fifties.

If movies are one of your great interests, then check out the Internet Movie Database (http://uk.imdb.com) search directory.

ShopSmart (www.shopsmart.com) lists and reviews online retailers. The **Internet Movie Database** (http://uk.imdb.com) does what its name suggests and a lot more besides.

Increasingly search engines, such as **AltaVista** (www.altavista.com), allow you to search for just image, video or audio files, as well as standard web pages. You can often specify the time-frame for your search, to weed out old documents or articles that aren't relevant any more.

There are also a number of so-called **meta-search engines** that search a number of directories simultaneously. This is a very useful development given that one engine's database can differ markedly from another's. Most engines can also rank the results of your search according to their relevance and the number of search words found.

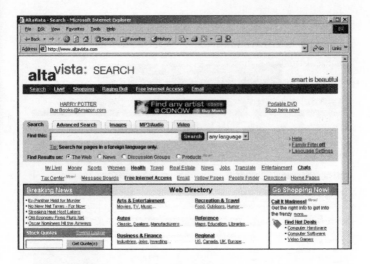

Two popular search engines, AltaVista (top, www.altavista.com) and Copernic (bottom, www.copernic.com).

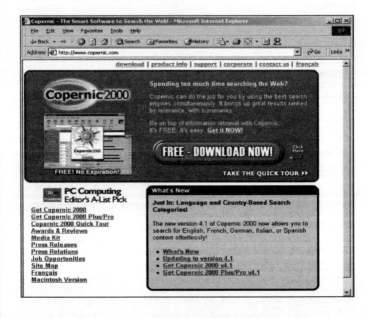

Search engines can also be used for finding public e-mail addresses, and there are plenty of electronic equivalents of the local business directory to help you find useful services – **Yellow Pages** (www.yell.co.uk), for example. There are also services, such as price comparison agents, that will scour the net for you looking for bargains.

But in short, there is still a long way to go before directories are sophisticated enough to be able to pinpoint surfers' requests speedily and accurately whilst remaining straightforward to use.

So what are the best search engines to use?

A lot depends on what you're looking for. If you're doing intense academic research, then you need a really comprehensive engine or specialist directory relevant to the work you're doing. An engine like **Copernic** (www.copernic.com) will trawl through at least 13 other search engines, directories, Usenet lists and e-mail databases looking for what you want. You need to download its browser software, which is closely integrated into Internet Explorer, but once you've done this you have an excellent search tool on your desktop capable of saving and categorizing your searches. Of course, searching such a large number of other engines does take longer, but it's generally worth it.

For very specific lists of sites grouped according to minority-interest categories, you could try **Webdata** (www.webdata.com), or **Super Seek** (www.super-seek.com). If you're just looking for good online shops to buy CDs from, you should be OK with the big name engines and directories (see pages 58–61). Much of the time it's a case of trial and error – which is why I don't like them very much!

There is a site called **Search Engine Watch** (www.searchenginewatch.com) that looks at all the latest developments in search engines, gives opinions on which are the best, and provides more tips on how to conduct accurate searches. Don't forget to bookmark your favourite search engines for ease of access. You could set up a dedicated 'Search' folder within your browser in which to keep them all. Browsers also have 'Search' tabs on their menu bars that take you to various search engines.

One search engine that has been winning plaudits for its speed and accuracy is **Google** (www.google.com). It does not have any distracting content of its own, just a plain vanilla search box. This helps increase the search speed,

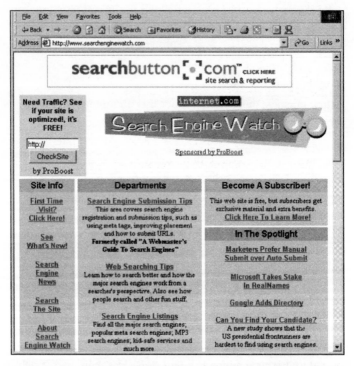

There are even websites to help you choose an appropriate search engine. Check out Search Engine Watch (www.searchenginewatch.com).

The simpler the better – by not having any content of its own the Google search engine is fast, accurate and easy to use (www.google.com).

which usually comes in at under 0.5 seconds. To give you some idea of the computing power involved, Google uses a mathematical equation involving over 400 million unknown variables capable of ranking all the pages listed and three billion terms. Now that's not something you can do with a pen and paper. Google has indexed over 250 million web pages – impressive, but still only a fraction of what's out there.

The really clever bit is that instead of searching absolutely every bit of text on a web page – a great deal of which is not relevant to your search – Google differentiates between important and unimportant sections, as indicated by the font size used, for example. It has learned to look at web pages with a discerning eye.

Anyway, here's a list of the main contenders...

Search engines

AltaVista	www.altavista.com
AltaVista UK	www.altavista.co.uk
Deja.com (for Usenet discussion groups)	
	www.deja.com
Excite	www.excite.com
Excite UK	www.excite.co.uk
FastSearch	www.alltheweb.com
Google	www.google.com
Goto	www.goto.com
HotBot	http://hotbot.lycos.com
InfoSeek	www.infoseek.com
Lycos	www.lycos.com

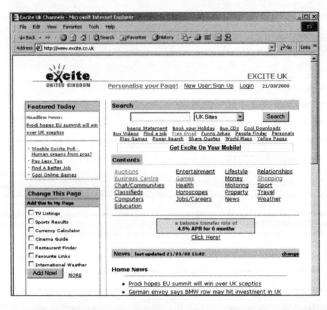

One of the most widely used search engines: Excite (www.excite.com).

Popular search directories include About.com (top, www.about.com) and Galaxy (bottom, www.galaxy.com).

Northern Light	www.northernlight.com
WebCrawler	http://webcrawler.com
Snap	www.snap.com

Search directories

Britannica	www.britannica.com
Yahoo!	www.yahoo.com
Yahoo! UK	http://uk.yahoo.com
About.com	www.about.com
DMOZ Open Directory Project	
	http://dmoz.org
The Argus Clearinghouse	www.clearinghouse.net
Galaxy	www.galaxy.com
Magellan	www.mckinley.com
EuroSeek	www.euroseek.com
LookSmart	www.looksmart.com
UKOnline	www.ukonline.com
What's Online	www.whatsonline.co.uk
UK Plus	www.ukplus.co.uk
Search UK	www.searchuk.com
UK Max	www.ukmax.com

Meta-search engines

Copernic	www.copernic.com
MetaCrawler	http://metacrawler.com
Dogpile	www.dogpile.com
MetaFind	www.metafind.com
ProFusion	www.profusion.com
All-In-One	www.allonesearch.com
Ask Jeeves	http://ask.co.uk
Powersearch	www.powersearch.com
Metasearch	www.metasearch.com

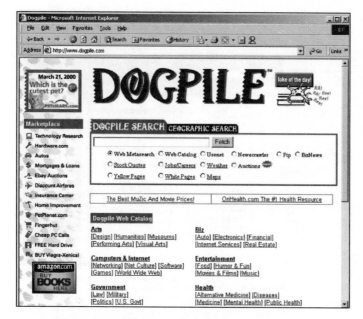

The metasearch engine Dogpile (www.dogpile.com).

Business and phone directories

Yellow Pages	**www.yell.co.uk**
Scoot	**www.scoot.com**
BT PhoneNet UK	
(directory enquiries)	**www.bt.com/phonenetuk**

Tracing e-mail addresses

Yahoo! PeopleSearch	**http://people.yahoo.com**
Bigfoot	**www.bigfoot.com**
WhoWhere	**www.whowhere.com**
Internet Address Finder	**www.iaf.net**

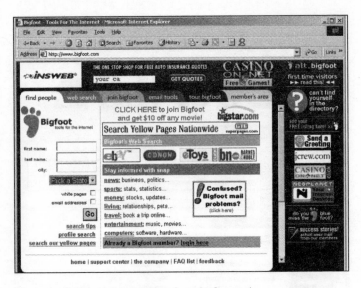

Trace e-mail addresses with Bigfoot (www.bigfoot.com).

Shopping agents

Search agents that can scan across scores of online retailers
looking for the lowest prices for products are the future of
the web. You can usually search using a number of criteria,
be they price or product type. Shopping agents give
unprecedented power to consumers and enable us to shop
globally. Making prices so easy to compare between retailers
is a sure way to stimulate competition and to keep prices
low. Retailers are finding that on the web they can no longer
hide behind their expensively bought brand awareness.
Shopping agents have blown all that out of the water – the
web is eroding brand loyalty with the help of these agents.

The US is ahead of the game as usual, with shopping
agents such as:

MySimon	**www.mysimon.com**
Bottomdollar	**www.bottomdollar.com**
Shopper.com	**www.shopper.com**

MySimon (www.mysimon.com) is a specialized search engine that can scan across online retailers looking for the lowest prices.

| Virtual Outlet | **http://vo.infospace.com** |
| PriceScan | **www.pricescan.com** |

Obviously there are important issues arising from shopping abroad, and these will be dealt with in a subsequent guide. Europe is beginning to cotton on to the immense potential of shopping agents. **DealPilot** (**www.dealpilot.com**) specializes in books, videos and music and is especially useful because it will scan US and European retailers, converting prices into your native currency, whilst including shipping costs as well. It will also give you the average delivery times quoted by the retailers.

The company also offers DealPilot Express – free software that you download on to your desktop. It only takes about a minute. When you're looking at a particular website,

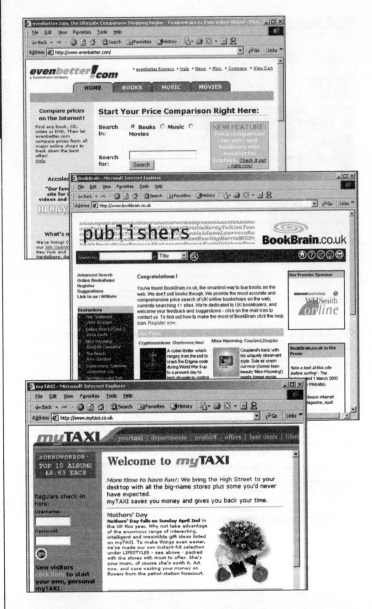

Specialist shopping agents include EvenBetter (www.evenbetter.com), BookBrain (www.bookbrain.co.uk) and MyTaxi (www.mytaxi.co.uk)

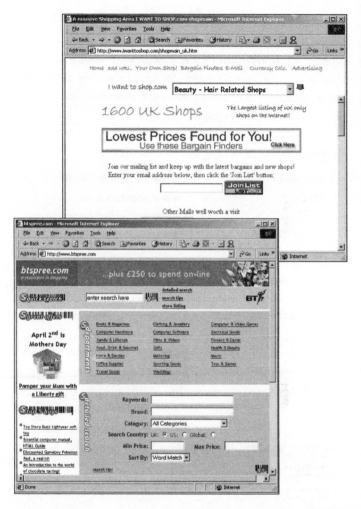

*I Want to Shop (top, www.iwanttoshop.com) and BTSpree (bottom,
www.btspree.com) are general-purpose shopping agents.*

DealPilot Express will automatically search the web in the background for the best prices for whatever it is you're looking for. **Bookbrain** (www.bookbrain.co.uk) is also very good for books, and UK-based **MyTaxi** (www.mytaxi.co.uk) has a slightly broader scope, including computers as well as videos and CDs, for example.

General purpose shopping agents include:

ShopSmart	**www.shopsmart.com**
BTSpree	**www.btspree.com**
Valuemad	**www.valuemad.com**
I Want To Shop	**www.iwanttoshop.com**

One thing is for sure: we are going to see a lot more of these agents cropping up, and they're going to become more and more sophisticated. A shopping agent is only as good as the number of retailers it can scan. As agents add retailers to their lists, the quality of their comparisons will become even more valuable. Soon we may see agents that can also take consumers' experiences into account, rating sites for reliability and speed of delivery, quality of service, as well as cost. With luck, we'll also see agents that can calculate total cost, including local taxes, import duty, Value Added Tax and any post office handling charges, opening up a truly global marketplace.

Search rules and tips

Much of the frustration associated with the use of search engines is caused by a lack of understanding about how to search. This isn't surprising as the rules are often very complicated and differ from site to site. But if you do intend to use the net a lot – and believe me it can become addictive – taking the trouble to learn a few basic search skills pays dividends in the long run.

Logic underpins most search engines. Words such as AND, OR, and NOT, whether spelled out or assumed, help computers sift vast amounts of data. These words are called **Boolean operators** after British mathematician George Boole.

For example, if you typed in 'dog AND bone' the search engine would find all documents where both those words were mentioned. If you wrote 'dog OR bone', the search engine would find all documents containing at least one of the words or both. This would result in a much larger, and probably unmanageable, number of documents. Searching on 'dog NOT bone' would have thrown up documents containing the word dog but not the word bone. That's Boolean logic. Other useful

> **TIP**
>
> *Use Boolean operators to narrow down your search.*

words include NEAR, which ensures that the words you've chosen are in close proximity and not scattered at opposite ends of a document with no relevance to each other at all.

Brackets round search words like this: (search phrase) can also tell the engine that you want to search for the phrase within the brackets exactly as written, rather than each of the component words. Again, each engine will have its own version of this, so read up on the rules. In quite a few engines speech marks – "search phrase" – will do the same thing.

These days many search engines use their own versions of Boolean searching, assuming you want all your search words to be found, rather than just any; other search engines assume you want any of the words. To confuse things further, some search engines use Boolean logic but denoted by different symbols. For example + and – before words can equate to AND and NOT. The treatment of upper and lower case can also be quirky.

It really is worth the effort finding out in detail the search system that a search engine or directory uses by investigating the 'Search Tips' section, usually found near the search box, or by choosing the 'Advance Search' option. It may be uninspiring to have to do this, but it will save you time in the

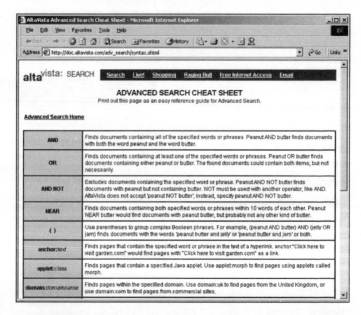

Learning how to use Boolean operators can save you many hours of wasted time on the net.

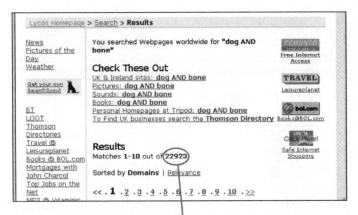

Searching on the web is easier said than done. A simple query can throw up tens of thousands of potential websites.

long run. There are search engines, such as Ask Jeeves, that attempt to answer questions you type into the search box. Although this is a laudable attempt to make search engines more intuitive to use, the answers you get are often completely unrelated to your question. This can even be more frustrating because the 'human' touch raises expectations of a more human response.

Berkeley, the famous US university, has a useful table on its site comparing the search features of some of the engines it recommends to its students. It is a very useful resource. Point your browser at this unwieldy address: **http://www.lib.berkeley.edu/TeachingLib/Guides/Internet/ ToolsTables.html.**

Still can't find what you're looking for?

The web is changing all the time and developing at a remarkable rate. Websites come and go and links in directories can become defunct. You'll often receive error messages telling you that the browser couldn't find the website, or the specific page you were after within a website.

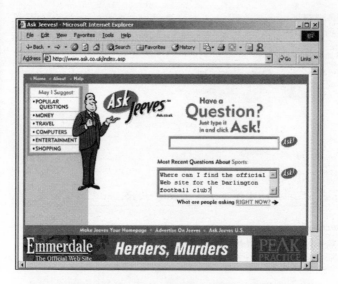

*Advanced search engines such as Ask Jeeves (above, http://ask.co.uk)
attempt to answer questions. If you want to know more about search
engines in general, check out the website at Berkeley (below,
www.lib.berkeley.edu/TeachingLib/Guides/Internet.html).*

TIPS

Here are some general tips to make searching less of a headache:

- *Try to use unique words and phrases*

- *Make sure you spell the words correctly*

- *Use several words rather than just a couple to help narrow the field down*

- *Learn how the search engine handles upper and lower case*

- *Use Boolean operators where relevant (or the search engine's own version)*

- *Compare results from several search engines or use a meta-search engine*

- *Try other sources of information within search engines, such as Usenet newsgroups, as well as the web*

- *When you're browsing all the 'hits' thrown up by a search, open new browser windows when clicking on the ones you're most interested in rather than reading each entry one by one and then pressing the back button to the results page. You'll save a lot of time this way.*

There can be a number of reasons for receiving error messages:

● **The website doesn't exist any more** – there's not much you can do about this. Sometimes you do get through to the website only to be told that it no longer exists, which is at least a little more helpful. If you're lucky you may be given a new link to go to. Directories worth their salt keep their databases up-to-date – there's nothing more annoying than a whole page of out-of-date links.

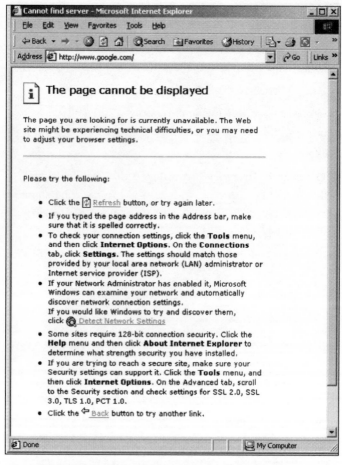

One of the most frustrating windows you can get on the internet!

②The page has been removed or updated – the information you're after may have been moved to another part of the site. On the error message page you'll often be given the main home page web address of the site that hosted the original page. If you click on this link you can at least scour the site for the information you were after.

Alternatively you can delete part of the web address in the address box of your browser until you reach the main domain name e.g. **www.sunday-times.co.uk** and then hit return. This will take you to the website's home page, too.

● **You typed the address incorrectly** – you may have forgotten the dot between the www and the domain name, or misspelled the website's name. Have a look and try again.

● **The search directory has indexed the site wrongly** – this should be a rare occurrence. You can test it by altering the web address slightly in a number of permutations. If you eventually get through and the directory is at fault, you should let them know in the strongest possible terms!

Chapter 4

Downloading Software

Plug-ins

Surfing the web is now a rich multimedia experience incorporating sound, images, video and animation. To make the most of all that the web can offer you, you need the right software that will let you listen to music, for example, or watch videos. These software programs are called **plug-ins**.

Software companies are competing madly with each other to make their versions of these plug-ins the definitive version. There are no standards, so that if a video file, say, is in a format devised by one particular software company, you have to download the plug-in that can read those files. Video files in another format will require another plug-in. There are usually several programs available that do much the same thing. This can be confusing, but at least it gives you the opportunity to try out a few and see which one you like best.

The latest browsers have several essential plug-ins already built in, but you don't have to stick with these. You can download alternative programs from the software companies' websites. (Again, you can find a number of these plug-ins on CD-ROMS that come with computer and net magazines.) A popular program for music and video, and the one that seems

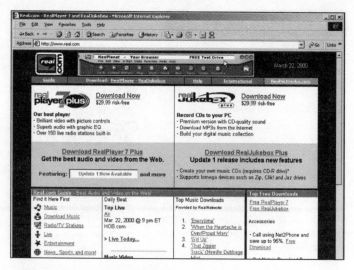

RealPlayer (above, www.real.com) is the leading plug-in for music and video. Apple's version is QuickTime (opposite, www.apple.com/quicktime).

to be dominating the market, is **RealPlayer** (**www.real.com**), which includes **RealAudio** and **RealVideo**. Microsoft has its own version called Windows **Media Player** incorporated into Internet Explorer. Apple's version is called **QuickTime** (**www.apple.com/quicktime**).

Such plug-ins are helping to transform the music industry. You can now listen to samples of songs online – the net's version of the traditional listening room – to help you decide whether you want to buy the CD. And you can even download entire CDs on to your hard drive. For more on downloading digital music, and alternative music plug-ins, *see* **Making the Most of the Web**, *page 129*.

Another important program is **Shockwave Director** and **Flash** (**www.shockwave.com**), which enables your browser to handle animated graphics and other advanced website design features. Many websites are using this program these days, so if you don't have it, your browser won't be able to

load the page properly. Usually, if this happens, the website provides a link to the software company so that you can download the required software there and then. But be warned: some of these programs are several Megabytes in size and can take a long time to download, especially with a slow modem (28.8kbps) and at a busy time of the day.

A lot of documents on the internet are now designed using **Adobe Acrobat** (**www.adobe.com**), which helps make pages look exactly as they would in a conventional book or magazine. But you need the **Acrobat Reader** plug-in to read these files (called **PDF files**). It is well worth getting, especially if you plan to print off documents – the quality is excellent. Also, websites often have forms that you

> ## TIP
>
> *Download software early in the morning (when most Americans are still in bed), or look out for free CD-ROMS accompanying internet and PC magazines. These often have lots of useful plug-ins on them and it takes a fraction of the time loading them on to your hard drive.*

Adobe's Acrobat Reader is one of the most useful plug-ins you can have.

need to fill in. Rather than completing them online, you can print them off before posting.

ActiveX controls

Other types of helper programs are called **ActiveX controls**, which start working as soon as they are downloaded. With most plug-ins you have to save them to a disk first, install them, then restart your computer. ActiveX controls are nimbler and have more scope. If you come across a website that requires an ActiveX control for you to read it properly, it will check your hard drive to see whether you already have it installed. If not, it will load itself on to your computer with your approval.

As ActiveX is a Microsoft invention, it is designed for use with the Internet Explorer browser. If you use a Windows 95/98/NT compatible version of Netscape you can configure the browser to accept ActiveX controls by downloading a plug-in called NCompass ScriptActive (**www.ncompasslabs. com**).

Java applets

Java is a programming language designed by software
company Sun Microsystems to work on any computer
regardless of the operating system (e.g. Windows 98 or Mac
OS). Websites can sometimes use mini-programs called
applets which load on to your computer as and when they
are needed. For example, if you launch some online services
you'll see 'Loading applet...' in the status bar at the bottom of
the page. The applet doesn't remain on your hard drive when
you go offline, so it doesn't use up disk space.

Some websites are also designed using an addition to
HTML called JavaScript, which facilitates animated graphics
amongst other things. The main problem with JavaScript is
that older browser versions have trouble reading it – pages
often don't load properly. You can either ignore websites
designed using JavaScript or better, upgrade your browser.

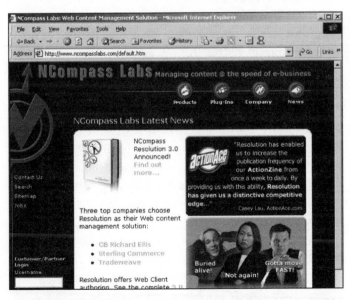

*NCompass ScriptActive (www.ncompasslabs. com) allows you to use ActiveX
with a Netscape browser.*

Free software

The net thrives in a culture of shared knowledge and experience. Consequently there are hundreds of sites that offer software that you can download for free. These may be enhancements to existing programs, or full programs in their own right. **Freeware** is, as its name suggests, entirely free. **Shareware** is usually free for an evaluation period, giving you the chance to try out the product before committing yourself. At the end of the evaluation period you're expected to cough up a registration fee.

It is common for shareware programs to have only a few out of many possible features activated. You can get a feel for the product and then decide either to stick with the limited free version or pay to receive the full version. So when you see lots of 'free' software advertised on computer magazine CD-ROMS, bear this in mind.

Another common practice is for software companies to release pilot versions of their products, known as **beta programs**. But be warned: these are not the finished article and often contain glitches that can interfere with other programs on your system. Net users are used as guinea pigs in effect.

The advantage is that you get to use the very latest products that could enhance your online experience greatly. If you do come across recurring errors, tell the software company and ask if there's a way to put it right. This is why they release such programs for free – to get feedback from net users and put things right before they sell them commercially.

Often correcting errors is a simple case of downloading a line of code. An enhancement or addition to an existing program is called a **fix** or **patch**. Not many software programs are entirely glitch-free, and all can be improved. So again, it's up to you to scour software company websites on a regular basis for news of improvements to your existing programs. It's a bit like fine-tuning your car to keep it in tip-top condition.

Most programs you download will have a **.exe** extension. When you click on the file in the directory you downloaded it to, the program will unravel itself automatically. But some files you download may be in other formats. A common one is **Zip** which has a **.zip** extension. These are compressed files that don't take as long to download. To unpack these compressed files you'll need an expander program, such as **WinZip**, **NetZip**, **PKZip**, or **Stuffit Expander**.

Some useful and popular software sites include:

Download.com	**www.download.com**
Shareware.com	**www.shareware.com**
Tucows	**www.tucows.com**
WinFiles	**www.winfiles.com**
Stroud's	**http://cws.internet.com**
Filez	**www.filez.com**
FTP Search	**www.ftpsearch.com**

Tips on downloading

The first thing to say about downloading software is that it can take a very long time if the program is several Megabytes in size and you only have a 28.8kbps modem. Deciding to download the latest version of your favourite web browser in the middle of the afternoon – usually the busiest and slowest time on the web – is not the best strategy. It is also the most expensive time to do it, even though you're usually only paying for a local rate call.

If cost is the main consideration it makes sense to wait until the evening before you start downloading, when phone rates are lower. If speed of download is your main concern, try to do it very early in the morning, when most Americans are asleep. If you're downloading from a UK site, you should

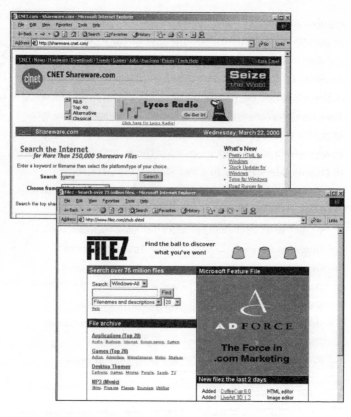

Sharing software is a long-held tradition on the net and there are scores of sites from which to download.

find downloads happen a little quicker after midnight, when most of us are in bed.

You often get a better distribution speed from a dedicated FTP site, which does nothing but supply software. Servers that allow you to access them and transfer FTP files are known as anonymous FTP servers. Generally, you log in first, then browse directories and subdirectories full of files, clicking on those you want to download. These files can often have very intimidating titles consisting of a seemingly random jumble of letters and numbers, so it's very important to get the title of the file exactly right.

You'll often find that ISP or PC helpline technicians will recommend that you download a specific patch or fix to iron out glitches that occur from time to time with your system. These little programs and extra bits of code are the most likely to have uninspiring names. Be sure to write down the exact filename on a piece of paper before you go looking for it. Downloading the wrong file could have unhappy consequences for your PC, especially if it's an update for the operating system, for example.

Bear in mind that when you are downloading software you don't have to sit there watching the progress box, you can send e-mail and surf at the same time. Get into the habit of opening new browser windows. And before you go mad downloading every bit of available software on the net, consider the effect on your hard drive memory. If you have a fairly old computer, space can be used up remarkably quickly.

FTP clients

There's nothing more annoying than for your connection to break down when you're most of the way through a big, time-consuming download. You may also have to dash out, either leaving your PC online (expensive) or aborting the download altogether (maddening). More often than not you have to start all over again.

TIP

When you decide to download a program or file, you're given a choice of whether to 'save it to disk' or 'run it from its current location'. It's better to save it to your hard drive — you choose which folder you want it to go in. Some browser versions will automatically place such files in a 'Download' folder.

But there are special **FTP programs** or **clients** that help to manage your software downloads. They can stop a download half way through and begin again at a later time at the place where they left off. They can also scour the web for the best and quickest FTP sites, allow you to schedule times when you want to download and dial your modem when you're not there. When downloads are completed they can even shut down your computer for you. Some advanced versions can chop up a program into parts and download those parts from separate servers to speed up the process. The only snag is that not all FTP servers allow downloads to be resumed part of the way through. Still, if you plan to download a lot of software an FTP program is a good idea. And many of them are free. Here are some suggestions:

WARNING

When you download programs from the web or from CD-ROMS you are letting something invade your computer that could contain bugs and viruses. A good anti-virus package is essential (see **Safe Surfing**, *page 159) and you should scan all programs first before installing them. Some FTP programs will automatically activate your anti-virus software after downloading.*

GetRight	**www.getright.com**
CuteFTP	**www.cuteftp.com**
Go!Zilla	**www.gozilla.com**
NetVampire	**www.netvampire.com**
Fetch (for Macs)	**www.dartmouth.edu/ pages/softdev/fetch.html**

Programs that help you manage your software downloads can save time and remove frustration.

Chapter 5

Faster Surfing

Introduction – the need for speed

Not for nothing has the web been dubbed the World Wide Wait. Broadly speaking, as it is a network of networks it is only as fast as the slowest link in the chain. You could have the fastest modem and computer, but if there's a bottleneck anywhere along the line, you'll find yourself tapping your fingers on the desk in frustration.

Something strange happens when we go online. We become very impatient. The net's unquestioned ability to get us information faster than ever before has raised our expectations, perhaps to unreasonable levels. What was considered a near miracle just a few years ago is pretty much taken for granted now. A great deal of hype and misunderstanding about the net has led us to expect that we can get whatever we want whenever we want it. Technology has been struggling to keep up with our galloping expectations.

So-called early adopters – the net nerds and gizmo gurus who live for every new technological development – were more tolerant of the net's failings because they knew the extraordinary amount of computing genius that lay behind this communications phenomenon. The new wave of surfers, on the other hand, who don't really care how it works – and why should they? – want it all and they want it now! In survey after survey lack of download speed is cited as one of the major frustrations of using the net.

The problem is that as the web experience has become immeasurably richer with the addition of colour images, animation, streaming audio and video, all these extras have placed an even greater burden on the network. Image and sound files are generally much larger than simple text files. But there's only so much data you can fit down a pipe at any one time, although there are technologies around that seek to increase this by various means. Add to this increased volume of data the sheer number of new people coming online and it's no wonder the net seems to be creaking at the seams sometimes.

As fast as the technology companies that supply the net's infrastructure upgrade and improve their systems, more seems to be required of them. And the internet motorways, or **backbones**, may be the fastest fibre optic cables around, but if you've got a slow modem, vast hordes of data will come charging to your computer, only to be forced to enter in single file. Most new computers now come with a 56kbps (56 kilobits per second) internal modem as standard. That's a darn sight better than 28.8kbps, but it's still nowhere near what we need to fulfil the net's potential.

Sufficient **bandwidth** is the net industry's Holy Grail – the greater the bandwidth, the more data you can fit down the pipe. At the moment, live video pictures are still rather jerky because they require so much data to be transferred at once: most computers with standard connections simply can't cope.

Luckily there are some new technologies just around the corner that should transform our collective web experience.

The search for greater bandwidth is never ending.

They are so-called **broadband services** that can transfer data at up to a hundred times faster than the fastest modem. This level of service seems to have been tantalizingly close for ages. But it looks like we may finally enjoy the benefits of high-speed connections soon.

DSL – Digital Subscriber Line

There are several DSL technologies around but they all involve sending digital data down conventional telephone wires at high speed. The most common form, and the one we'll be sold by the telecoms companies – British Telecom in particular – is **Asynchronous Digital Subscriber Line** or **ADSL**.

This technology allows data to be transferred across the telephone network at up to 2Mbps (two megabits per second), 35 times faster than the fastest 56kbps modem. The asynchronous bit refers to the fact that downloading is faster than **uploading**. As most net activity involves downloading rather than uploading, this doesn't matter much.

The general public is likely to be sold a slower service of up to 512 kilobits per second, but still nearly ten times faster than what we have at the moment. Some industry analysts believe BT will restrict the speed for ordinary customers because it doesn't want its business customers – who can pay around £30,000 a month for a 2Mbps permanent connection called a **leased line** – to migrate in droves to this cheaper service if a similar speed is on offer. At the time of writing ADSL was thought likely to cost around £50 a month for households, with connection charges on top.

ADSL offers many advantages. For a start, the line is always open and it's unmetered – you don't pay by the minute. This gets rid of all that logging on and off business and stops you worrying about the length of the call while you surf. It also means you can download software and memory-munching pictures and sound far more quickly, taking a lot of the waiting out of the web experience. When people send you e-mail you'll know about it instantly.

A faster, more efficient service will undoubtedly encourage more people to surf the net and for longer periods, and this will boost e-commerce. But without sufficient competition in the broadband market, it looks likely that we'll pay handsomely for this level of service for some time to come. At the moment there isn't enough financial incentive for BT to make ADSL *the* way to access the net, partly because of the ludicrously complicated charging structure telecoms companies are bound by for the use of each other's networks.

The UK is already one of the most expensive European countries for net access, even after the free-ISP revolution. For example, in Holland surfers can enjoy a 1.5Mbps unmetered connection for around £26 a month. This situation looks likely to continue unless Oftel, the telecoms regulator, does something drastic.

Cable modems

The UK cable television companies have been just as slow off the mark in introducing high-speed connections to the domestic market. Cable modems don't use the traditional telephone network but send data directly down fibre optic cables. NTL, the UK's largest cable operator, is currently rolling out its HiSpeed Internet service in its franchise areas, offering download speeds of up to 512kbps and upload speeds up to 128kbps.

At the time of writing it cost £40 a month, but that covers everything. As the system doesn't use the voice network there are no telephone calls to pay for and your connection is permanently switched on, giving you 24-hour unmetered access to the net. You can use the telephone at the same time. The only snag – and there's always a snag – is that you need to buy a cable modem starting at around £150 and you need an Ethernet card (10BaseT) fitted in your PC. You also need at least 60MB of free hard disk space and 32MB RAM. At the time of writing NTL's service is only compatible with Macs running OS 8.5. NTL is offering to install all the equipment and set up your computer for around £200.

Nobody said high-speed access was going to be cheap. The cable companies are likely to roll their net services in with their cable television and telephone services in a variety of marketing packages.

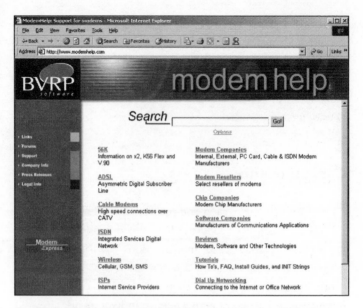

The Modem Help site (www.modemhelp.com) has excellent advice for squeezing the most speed out of your modem.

ISDN – Integrated Services Digital Network

BT's forerunner to ADSL is ISDN, another higher-speed net connection enabling customers to reach speeds of up to 128kbps. It was mainly designed for the business market, but BT has packaged it for the home market as Home Highway. It dispenses with the need for modems but does involve the installation of a terminal adaptor at both ends of the line. You get one normal phone connection and two ISDN connections. To get the maximum 128kbps speed you have to combine the two ISDN lines.

Again, at around £27 a month plus installation costs of over £100, the service isn't cheap. And with the promise of the much faster ADSL just around the corner, it's no wonder that ISDN hasn't taken off in a big way in the domestic market.

Tips for faster surfing

While you contemplate which high-speed service to start saving up for, you can at least tune your PC so that it can make the most of what it's got. Here are some tips for achieving faster surfing:

❶ Always buy the fastest modem you can find (56kbps) – you'll more than recoup the cost (around £100) as faster downloads cut down your time online.

❷ Squeeze the most speed out of your modem by following the advice at the excellent **Modem Help** website (**www.modemhelp.com**).

❸ Use a Net accelerator program that can reconfigure your operating system to download data in the most efficient way and allow you to browse pages offline. Some suggested programs:

Surf Express	**www.connectix.com**
Speed Doubler (for Macs)	**www.connectix.com**
Accelerate	**www.webroot.com**
SpeedNet	**www.paramagnus-development.com**
EasyMTU	**http://members.tripod.com/~EasyMTU**

❹ Learn how to use search engines effectively: it can save you a lot of time.

❺ Use the right mouse button to speed up your web browsing. Right-clicking on web pages and links calls up a neat time-saving context-sensitive menu.

❻ Take the time to customize your browser's start-up page so all the services and links you want are there ready for you when you log on.

You can accelerate download times using programs from companies such as these.

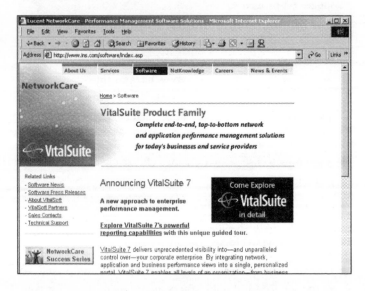

❼ Use a diagnostic tool such as NetMedic from **VitalSigns** (www.vitalsigns.com) to spot log-jams at various points on the network. It could be your modem slowing things up, in which case upgrade. It may be you ISP being inefficient – consider switching to another.

❽ Make sure the serial port setting is at the right speed. For a 56kbps modem the speed should be set to 115,200bps. In Windows 95/98, go to 'My Computer', 'Control Panel', 'Modems', 'Properties' or for Macs, go to the 'MacPPP' control panel.

❾ If you're experiencing slow connection times ask your telephone company if the gain on the line is at the optimum level for data transfer.

❿ Get into the habit of reading and composing your e-mail offline. It's surprising how easy it is to forget you're still online. You can tell your e-mail program to disconnect once it's downloaded all the messages.

⓫ Do your surfing at quiet times of the day. If you're looking at US sites in particular, get online in the morning when most Americans are asleep. Things can get busy and slow in the afternoons and evenings.

⓬ If it's just text-based information you're after choose a 'text only' option if one is offered as the page will download much quicker without the graphics.

⓭ Tell your browser not to bother loading images, sound and video. Go to 'Tools', 'Internet Options', 'Advanced' in the latest version of Internet Explorer, and 'Edit', 'Preferences', 'Advanced' in Netscape Navigator.

⓮ Check the website of your modem manufacturer for any news of upgrades for the software used to run the modem (known as firmware) and also the drivers that tell your PC how to communicate with the modem.

🚯 I've said this elsewhere but it's worth saying again: don't keep clicking backwards and forwards between pages, just open new browser windows so that you can view several pages at once. It saves a lot of time.

🚯 Close other programs that you don't need open when you're surfing. This will free up more of your computer's RAM to concentrate on downloading. Ideally you should have 32MB RAM at the bare minimum these days. Highly specified computers will have 128MB.

🚯 Consider using a simpler browser, such as Opera, which doesn't take up as much memory as the other popular browsers. This also helps speed things up.

Cacheing web pages

When you browse the web your computer stores the pages you visit in a special folder on your hard drive known as your **cache** or **temporary internet files**. When you call up a web page in your browser which you have visited before, your computer simply takes it from its own cache file rather than wasting time downloading it from the web. This means you can also browse through these files when you're offline.

One way to speed up your browsing is to increase the amount of disk space allocated to your cache folder. Be careful though, the web page may have been updated since you last visited it, so you need to press 'Reload' or 'Refresh' in your

> **TIP**
>
> *If you're accessing a page that is regularly updated (with news or financial data, for example), remember that you may be viewing an old page cached on your computer. Update the page using the **Reload** or **Refresh** button.*

browser to update the page. This is especially important when visiting financial data sites where information can change every few minutes. If you allocate too much memory to the folder, it can become unwieldy and slow your whole system down – the opposite of what you're trying to achieve. The right size cache depends on the amount of RAM you have and hard disk space. Ask your ISP for advice on the best settings for your computer.

Internet Service Providers often store copies of commonly accessed web pages on specially assigned servers called **web proxy caches**. These can save time by intercepting your request before it has even reached the ISP's server. This means that rather than going directly to a website that could be stored halfway round the world, your request might just stay relatively local. The shorter the distance, the faster the speed of response.

Check whether your ISP uses web proxy cache and ask how to set up your computer to make the best use of it. To bypass a cached page and go direct to the website, simply press the shift button on your keyboard and click 'Reload' or 'Refresh'.

Keeping the telephone bill down

The 'faster surfing' tips above will make your online experience more enjoyable, but they can also help keep your telephone bills down. The theory is that the faster you surf and find what you want, the less time you'll spend online. That's all very well if you are after something specific, like a sports report or bank balance. But if you're online just for entertainment – listening to the latest CD samples at an online record store, say – your concept of time can become a little blurred. Surfing should carry a health warning: it can become addictive. And with the free ISP

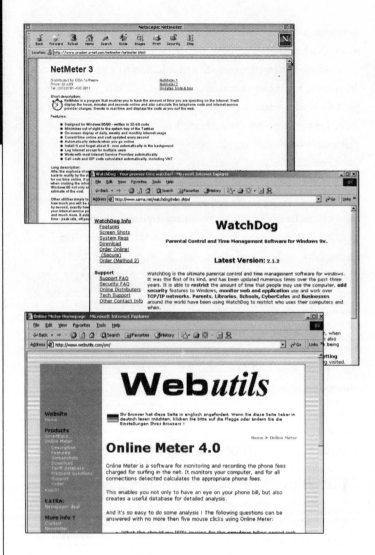

There are some very useful programs around that can help you monitor the time you've spent online, including NetMeter (www.cracker.u-net.com/netmeter/netmeter.html), WatchDog (www.sarna.net/watchdog) and Online Monitor (www.webutils.com/om).

revolution it can be tempting to think everything is for free. Until the phone bill arrives.

There are programs around that can help you monitor how long you've been online and how much it is likely to cost you. These are especially useful if other members of your family regularly surf the net and you want to keep a tight rein on the time spent online.

Try these out:

NetMeter	**www.cracker.u-net.com/ netmeter/netmeter.html**
Online Monitor	**www.webutils.com/om**
WatchDog	**www.sarna.net/watchdog**

The other thing you can do is shop around for a new telephone company. At last there does seem to be some real competition entering the market as the cable companies consolidate and take on British Telecom head on. It's quite complicated working out all the different tariffs, and the best option for you will largely depend on when during the day you do most of your surfing.

For help on working out what tariff would suit you best, try a very useful service called **Tariff On Line** (**www.toll.co.uk**) from TMA Ventures, the commercial arm of the Telecommunications Managers Association. It provides tariff calculators for terrestrial phone services *and* mobile phones.

Some cable companies offer free local calls in the evening and at weekends – ideal for light recreational surfers. And bear in mind that you don't have to subscribe to cable to benefit from alternatives to British Telecom. For example, dialling 132 before a telephone number will put you on the Cable & Wireless tariff for a monthly charge of around £1.25. You get local off-peak calls at 25% off the BT rate and you qualify for their free calls offer.

Another way to keep your own costs down is to surf while at work. It is very common although some companies are very strict about the amount of time you're allowed to spend online and

the type of files you download. Several employees have been sacked for viewing pornography and other unsuitable sites while at work. Companies that have their own internal computer networks will guard them from outside infiltration using security systems called **firewalls**. These will often filter the type of web pages that can be downloaded, especially those that require you to enter a user name and password for security. This means that you may not be able to access some services, such as banking or stockbroking sites, from work.

If you have a tolerant employer that doesn't mind employees spending some time online for private use, you could have a chat with the information technology manager to see if there's a way you can access your sites without the company's firewall rejecting them.

Unmetered calls

After intense pressure from the net industry and telecoms regulators, BT has finally agreed to offer fixed-price tariffs for unmetered, unlimited net access. At the time of writing, BT's proposed product, Surftime, is to cost £29.25 a month for 24-hour-a-day unlimited access. For those surfers who use the internet extensively and who have no choice but to surf during the day when calls are at their most expensive,

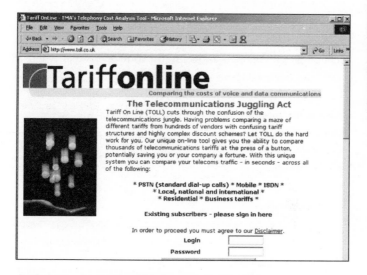

Tariff On Line (www.toll.co.uk) helps you calculate which telecommunications tariff will suit you best.

unmetered tariffs are a godsend and should save them a considerable amount of money. Cable companies, such as NTL and Telewest, and ISPs, such as Freeserve and Virgin Net, are following suit with a range of packaged offerings. Even search engines such as AltaVista are entering the unmetered calls market.

It's less clear where the savings lie for lighter users. There will be a range of tariffs, some restricting users to unlimited access just at the weekends, for example. There's no doubt that this move is a step in the right direction.

WARNING

Don't confuse unmetered calls with a high-speed 'always-on' net access. By offering unmetered calls BT isn't offering to make your internet connection speeds any faster.

Don't confuse unmetered calls with a high-speed 'always-on' net access. By offering unmetered calls BT isn't offering to make your connection speeds any *faster*. That's up to the alternative technologies mentioned above. It just so happens that when ADSL and cable modems do become widespread, they will also be unmetered. Soon the only cost we'll need to worry about is the monthly all-in charge for whatever way we decide to access the net.

Chapter 6

Electronic Mail

What is e-mail?

Electronic mail is a way of sending typed messages, documents and any other kind of digital files to other people with an internet connection and an e-mail account. When you sign up with an internet service provider, you are automatically given an e-mail address along the lines of 'yourname@ISPname.net'. You can often have several e-mail addresses. This is useful if several members of your family use e-mail – it helps messages to go to the people they were intended for.

What's so good about it?

It is easy to forget with all the emphasis on wacky websites with fancy graphics and daft content in the media, that e-mail is the net's most successful application. The net is all about communication after all. This relatively simple activity has revolutionized communications across the world. A recent survey found that 60% of US net users prefer reading their e-mails than reading ordinary post and 34% prefer to send an

e-mail than make a telephone call. They've even made a
Hollywood film about it.

The beauty and significance of e-mail is not so much the
ability to communicate instantly with people across the world
– the telephone can do this after all – it's the versatility that
comes with it. Along with your simple message you can
attach digital files that could be whole books, pictures, sound
files, or video snippets, and send them globally in a matter of
seconds. And now you can send e-mails written in HTML,
enabling you to incorporate graphics and links into the
message itself.

Yet you can do all this at the cost of a local call, because
you're still just paying for the link to your ISP. In the US,
where they mostly get local calls for free, it costs them
nothing. So you can send photos of the kids to relatives in
far-flung parts of the world very cheaply and quickly. And
when you send an e-mail to the other side of the globe, you
don't have to worry about waking people up, unlike the
telephone. They can just pick up the message in the morning
when you've gone to bed!

Another significant advantage of e-mail is that you can
send and receive messages wherever you happen to be, but
the people you're communicating with don't have to know
where you are. You can remain elusive yet stay in contact. It
puts you in control. The reason for this is that when someone
sends you an e-mail it doesn't go straight to your computer, it
goes to your ISP's computers first. The messages stay there in
a file reserved for your messages only. When you want to
retrieve them, you go online, fire up your e-mail software and
download them from your ISP's server to your computer or
whichever computer you happen to be using. So you can still
access your e-mail from a computer in Singapore if you like.

E-mail has also been very significant for businesses whose
postal costs are often astronomical. Now they can send the

same document to hundreds of recipients very easily using electronic mailing lists. There's no need to print out the document hundreds of times, put the copies in envelopes, attach stamps and find a postbox. It's all done swiftly and efficiently, saving paper, postage, storage and time.

The speed of delivery also makes it a much more efficient way of keeping in touch. Responses can be much more spontaneous and intimate even than written letters. The lack of formality of a hastily typed e-mail encourages frankness and honesty. If you have the chief executive's e-mail address, your message goes straight to the top. Before, if you wrote a letter, more than likely it went through an elaborate filtering process involving various flunkies whose sole job seemed to be preventing the boss from being disturbed. You'd be lucky if the boss even saw your letter.

And it works both ways. If you make something easier for people to do they will generally do it more often. If that chief executive thinks he can reply in just a few words that will take him half a minute to write, he's more likely to do so. Speaking personally, as an internet journalist for *The Sunday Times* I leave my e-mail address at the end of my weekly column. I get a lot of e-mails from company bosses who would probably not have bothered if they'd had to write a conventional letter. I also get plenty of useful e-mails from readers recounting their experiences. They probably wouldn't have bothered either unless they'd had e-mail.

The speed with which e-mail can be sent means that service providers can keep in touch with their customers far

TIP

All the messages you've sent and received can be archived in separate files, making them easy to retrieve and organize. It is very easy to add senders' e-mail addresses to your address book. It also saves on paper and storage space.

more easily. They can tell you when they are running special offers, for example, or when your favourite shares have risen or fallen in value. Software companies can tell you when upgrades for their products are available, including links in the message to the exact page on their website. If you sign up for newsletters and online magazines you can have them delivered to your in-box every day. E-mail can be far more reliable than the paper boy or girl, ready to deliver your messages as soon as you start your computer in the morning.

When easier isn't always better

E-mail is so easy to use it may land you in trouble if you're not careful. It can be very tempting to rattle off an e-mail to someone when your blood is up and say something you regret later on. At least with traditional letter-writing you had time to cool off and think better of it. With e-mail, once you've pressed that **Send** button there's no going back – you can't change your mind once that digital rocket has been despatched across the network. So be careful.

Writing quickly can also mean writing thoughtlessly. Phrases meaning one thing and intended to be read in a certain way, can often be misinterpreted by the reader. E-mail can make it easier for you to give offence unintentionally, too. So before sending your e-mails, check them over first and apply the normal rules of letter-writing. Some people feel that because a message is electronic they can dispense with common courtesies, such as 'Dear...', 'Yours sincerely' and so on. Some people don't even bother to include their names.

What do I need to set up e-mail?

First of all you need an e-mail address, normally assigned to you when you sign up with your ISP. Your address is normally based around your name before the @ symbol and the ISP's domain name after it. But don't assume that you'll be able to have whatever name you want. A lot of people with the same name as you may have signed up to the same ISP. If they've already bagged the name you wanted you'll normally find that you have to settle for a variation on it, usually involving the addition of some numbers.

Sometimes it can take a long time just entering proposed user names into the registration box only for the ISP's server to throw them back at you because they've already been taken. If the ISP is very popular you can sometimes end up with an address that looks more like the code for an Inland Revenue leaflet rather than your name.

You also need an e-mail program or **mailer** – a dedicated piece of client software designed to handle all your e-mail messages. The good news is that such programs now come bundled in the Navigator and Internet Explorer (IE) web browsers. Navigator's mailer is called **Messenger** and IE's is called **Outlook Express**, or **Outlook** in the more advanced version.

As there are several different versions of these packages around there's not space to go through each of them in detail. If some of the instructions given in this section don't exactly tally with your version, look up the key word in the 'Help' section of your browser program. Unlike other Help sections in various software packages, the ones in mailers are pretty good.

You can launch the e-mail program in IE simply by clicking on the 'Mail tab' in the browser's menu bar at the top. In Netscape choose 'Messenger' after clicking on the 'Communicator' menu at the top.

Outlook Express and its enhanced version Outlook will also place an icon on your desktop. This is quite handy because it means you can just launch the e-mail software by

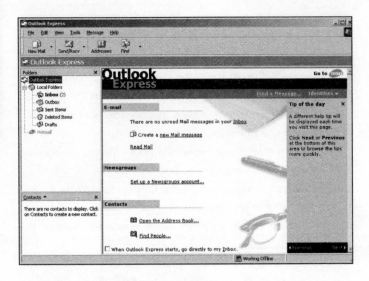

Two of the most popular e-mail programs: Outlook Express (above) and Eudora (opposite).

itself, not the web browser, if all you want to do is send and receive e-mail at that time.

Fortunately these two programs are quite sufficient for most people's needs, although **Eudora** (**www.eudora.com**) is a popular alternative. There are plenty of others on the market too. You can research them at sites such as:

Download.com	**www.download.com**
Winfiles.com	**www.winfiles.com**
Dave Central	**www.davecentral.com**

Configuring the software

Mailers are pretty similar in look to web browsers. They have their own tabs and pull-down menus. To make the most of the features they offer it's a good idea to spend time browsing through them and getting to know them.

When you first load your ISP software that contains your browser and e-mail program, a set-up 'wizard' –

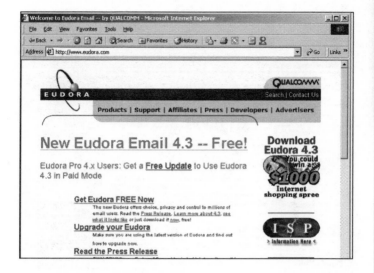

automated installation program – takes you through the process step by step.

The main task when setting up the software is putting in the names of your ISP's incoming and outgoing mail servers. Your ISP will tell you what to enter in these boxes. It's usually something like 'mail.ISPname.net'.

The outgoing mail server is also called the SMTP server, which stands for Simple Mail Transport Protocol, the agreed standard for sending mail across the net. The incoming server is usually a POP3 server, which means Post Office Protocol. POP3 allows you to access your e-mail from anywhere on the net, even if you're online with another ISP.

You also need to enter your account name – usually the first part of your e-mail address. A common mistake people make when setting up the software is to put their full name in the 'Account name' box. For example, although my name is 'Matthew Wall' the account name for my e-mail address is actually 'mt.wall'. If I get this wrong the server can't recognize the name and won't let me get at my mail.

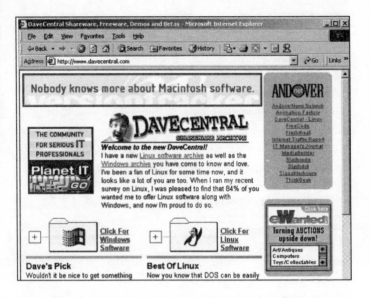

You can research alternative e-mail programs in sites such as Dave Central (www.davecentral.com).

Another important part of the set-up process is the password. You don't want people being able to get online and rifle through your mail do you? As with all passwords, make sure you keep it secret. If you forget it, tell your ISP and providing you can pass some other security checks, such as your mother's maiden name, date of birth and so on, they'll remind you. The latest dial-up connection software allows you to store passwords on the system so that you can connect automatically when you launch the software without having to fill in any boxes.

Obviously this is more convenient but less secure. Anyone with access to your computer could interrogate your e-mail in-box, read your mail, and worse still, potentially send rude, career-damaging messages to your employers. Not a suitable option for the workplace then.

Writing, sending and receiving mail

The great thing about e-mail is that you don't have to be online to write messages. You can compose them at your leisure and then send several messages all in one go, saving online time and telephone bills. Just find the 'New Message' tab or its equivalent, type in the e-mail address of the person you're writing to at the top of the page, give your message a subject in the 'Subject' box, write your message and press 'Send'. It's that simple.

In Outlook Express this will automatically file the message in the 'Outbox' ready for sending when you next go online. In Messenger you just choose 'Send Later' from the 'File' menu. You can close down your mailer programs and come back to them later if you want. Your unsent messages will still be there.

To send them, go online and press the 'Send/Receive' tab in Outlook Express, or 'File', 'Send Unsent Messages' in Messenger. The latest version of Messenger shows a pop-up box asking if you want to send the unsent messages. To receive messages you simply click on the 'Get Msg' tab.

In Outlook Express, clicking the 'Send/Receive' tab will also download messages into your 'Inbox.' If you don't want to receive messages yet, but just send them, you have to click on 'Tools' then 'Send'. This can be confusing because in the latest version of Outlook the drop-down menus only show those options you use most often. So the 'Send' option can be hidden. If you click on the chevrons at the bottom of the drop-down menu it reappears.

Once you've sent your messages they are automatically stored in a 'Sent' folder, which is a very useful resource. You can remind yourself what you actually said to someone before you write back at a future date. These sent messages can be sorted in several ways, too, including by name of recipient and date of sending.

When you receive messages you can read them straight away even while others are still downloading. Just double-click on the message and it will open in a new window. If you want to reply immediately click on the 'Reply' tab at the top of the window and the sender's address will automatically be entered into the address box. You normally write above the text that was sent to you, although some people prefer to edit the quoted text and put their replies underneath. This is such a fast way of communicating that it's possible to have a near instantaneous discussion with someone, where each part of the conversation is recorded in the body of text below. Unless you change the title in the 'Subject' box, your reply will contain the same title with 'Re:' in front of it.

You can also forward the message you've received simply by clicking on the 'Forward' tab and then entering the e-mail address of the person you want to forward the message to. You can tell if you've received a forwarded message if you see the abbreviation 'Fw:' or 'Fwd:' in the subject line.

Building up an address book

You can store e-mail addresses and other contact information, such as telephone numbers and job titles, in your address book. The quickest way to add addresses to the book is to right-click on the sender's address. This opens up a menu of options. In Outlook you choose 'Add to Contacts'. This then opens an address book page giving you the option to add other relevant contact details and index the person's name in the way you want.

In Messenger choose 'Add Sender to Address Book' and you're given a similar page to fill in. You save these and that contact is then added to the book. The advantage of this is that it saves time but also enables you to type just the name of the recipient in the address box, rather than the whole e-mail address. The e-mail program recognizes the name and supplies the e-mail address for you – providing you type the name in the same way you logged it in your address book.

> **TIP**
>
> *You can store e-mail addresses and other contact information, such as telephone numbers and job titles, in your address book.*

To add an e-mail address to the book from scratch in Outlook, go to 'Tools', 'Address Book', 'New', then 'Contact', or just click on the book icon in your e-mail browser menu. In Messenger, 'Address Book' is found in the 'Communicator' section. It calls new contacts 'Cards'.

You can start a new message to one of your contacts in the address book in a number of ways – from inside or outside the address book. For example, in Messenger if you right-click on the contact name then choose 'New Message', the address is automatically inserted into the address box of the new message. In the Outlook address book, right-click on the chosen address, choose 'Action', then 'Send Mail'.

Back it up!

Over time your e-mail address book can become a significant and valuable resource. Don't run the risk of losing it if your e-mail software becomes corrupted for some reason. Save it to a floppy disk just in case. In Outlook, choose 'Import and Export' from the file menu, then the 'Export to a file' option. You then have to select the format you want the file to be saved in – 'comma separated values (Windows)', for example. This doesn't much matter so long as you remember the file type you chose when you want to import the file back into your browser. If you get them mixed up you're likely to end up with a scrambled address book that's no use to anybody.

It does matter if you want to access the file from outside your e-mail program. Then the correct file format will depend on the software you have on your system best suited to reading these kinds of databases. A little trial and error is usually called for. But there's nothing to stop you saving the file in a variety of formats and seeing which one works best with the database software you have.

You then have to select the folder you want to export. In Outlook the address book is the 'Contacts' folder. Click on it then choose a destination for the file – a:\Contacts Backup, for example, and then export the file. To import the file, you just carry out the same procedure in reverse.

The process is a little simpler in Messenger. Open the address book then choose 'File', 'Export'.

> **TIP**
>
> *Over time your e-mail address book can become a significant and valuable resource. Don't run the risk of losing it if your e-mail software becomes corrupted for some reason. Save it to a floppy disk just in case.*

Choose a destination drive and name for the file, plus a format, then click on 'Save'.

Sending mail to several people at once

Once you've chosen the main recipient of the e-mail and put the address in the 'To:' box, you can send the message to others by adding their addresses in the 'CC' (carbon copy) box underneath. You can put as many addresses in here as you like, choosing from the address book, or typing them in manually and separating each using a semi-colon. The main recipient will be able to see the addresses of all the people you've copied the message to.

If you don't want the recipient to see who else you've sent the message to, put their addresses in the 'BCC' (blind carbon copy) box. This is especially useful if you've set up large groups of e-mail addresses and you want to send a circular to all the members of a club, for example. There's no point clogging up the e-mail header with lots and lots of addresses unnecessarily. But

> ### WARNING
>
> *Bear in mind that if you receive a message as one recipient on a list, if you choose Reply To All your message to the sender will also go to everyone on the list. The ease of doing this in a hurry has led to some spectacular bloomers with private and sometimes indiscreet messages being broadcast to entire lists of people.*

everyone on the list can still see the name and address of the main recipient and anyone else in the CC box.

To prevent disclosure of anyone on the list you can put your own e-mail address in the 'To:' box and put the list in the BCC box. This can be a little confusing for some people because they'll see that the message is to and from the same person! At first it looks like a mistake ... but hopefully the contents of the message should make it clear that the message is for them.

Creating e-mail lists

Your e-mail package can also help you to set up mailing lists very easily, which saves a lot of time if you're often sending messages to the same group of people. In Outlook just click on 'File', 'New', 'Distribution List'. Give the list a name then select the people you want on that list from your address book, or add them manually. In Messenger click on 'Communicator', 'Address Book', 'New List' and do the same.

Including a signature file

If you write a lot of e-mails you can include contact details with your message automatically every time you start a new one. It's a bit like attaching a business card and saves you typing out your details every time. It's called a **signature file**. In Outlook, start a new message then click on 'Tools', 'Options', then 'E-mail options'. Give your signature file a name then enter the details you want to include in the box provided. If you're clever you can create the file in a graphics package, adding colour and changing fonts, then import that file as your signature. But bear in mind that the more graphics you use, the longer it will take to send your messages. You also run the risk of alienating your recipients if it takes ages to download your fancy graphics file each time!

TIP

If you write a lot of e-mails you can include contact details with your message automatically every time you start a new one. It's a bit like attaching a business card and saves you typing out your details every time. It's called a signature file.

In Messenger click on 'Edit', 'Preferences', 'Mail' and 'Newsgroups' in the file tree on the left, then 'Identity'.

Messenger calls these signature files personal cards or vCards. You can include as much or as little detail as you like.

Sending attachments

When you compose an e-mail you can also attach other documents to your message, such as word processing files, spreadsheets, images and sound files. It's very easy. All you do is click on the paperclip icon in the mailer menu bar and then select the file you want to attach from whichever directory you've put it in.

When the recipient reads your message the attached file is shown as an icon along with the message. Click on the icon and the file opens. The only major problem you might encounter is that your recipient doesn't have the necessary software to read the type of file you've attached. For example, lots more people are using Adobe Acrobat for desktop publishing. But if you don't have the Acrobat reader software you won't be able to open these documents (called PDF files).

If possible, send

TIP

Avoid sending attachments that are very large or in an unusual format. Many people resent having to wait ages for something to download – especially if they then can't read it.

attachments in a common format, otherwise you risk annoying people. You also risk annoying people if your attachments are very big – more than a few hundred kilobytes say. They take a very long time to download and it's their phone bill you're adding to. It's the quickest way to make enemies, especially if your attachment isn't even of much interest.

Another problem, thankfully now uncommon, is that the e-mail program you are sending to may be using a different encoding standard – UUencode or MIME are the two main ones – for attachments. If your programs aren't in sync your recipient will just receive a lot of gibberish. You can specify which standard your package uses. For example in Messenger's 'new message' window, click on the 'Options' tab. You can then tick a box if you want to use UUencode instead of MIME for attachments. In Outlook, click on 'Tools', 'Options', then the 'Mail Format' tab. In the box next to 'Send' in this message format choose 'Plain Text', then 'Settings'. If your recipient is having trouble reading your attachments, experiment by switching standards to see which works best.

Web-style e-mail

The latest versions of e-mail programs are very sophisticated, incorporating many advanced word processing features, such as a wide choice of fonts, cutting and pasting, and other standard formatting facilities. You can even create new messages as **HTML** web pages capable of containing links to websites and including web graphics and sound files. If the recipient clicks on the link from within the message it will automatically launch the web browser software and, once online, go straight to the selected page.

This is great news for advertisers because they can send you promotional e-mail and give you the exact web address where you can buy whatever it is they're trying to sell. It's an effective form of direct marketing. It's good for us too because we can decide pretty quickly whether we want whatever is being advertised and delete it or act on it in a trice.

Rather than including attachments with your messages, HTML e-mail allows you to incorporate pictures, charts, photos, sound files and video snippets in the body of your message by clicking on 'Insert' in the main menu, then choosing 'Picture' or 'Object'. Once you've inserted the selected file, you can then format it, making it smaller or larger, for example.

Managing your mail

If you get a lot of mail that you want to keep, it can soon become a headache trying to keep track of it all. E-mail programs are excellent at helping you organize your messages. You can create new folders easily and simply 'drag and drop' messages into the folders you want. Sorting them by date received, size of file, or sender's name, is easily achieved by clicking on the tab at the head of each column within the folder.

You can set up your program to filter your mail into various folders if you want, rather than having them all arrive in your in-box. But this is only really worth doing if you receive tons of mail. Look under 'filters' in your program's 'Help' section.

You can mark your mail as 'read' or 'unread' and flag it up to remind you to read it later. And when you're sending messages you can specify whether they are high or low priority, and whether you would like a receipt to prove that they arrived.

E-mail for businesses

Businesses who do a lot of e-mailing to long lists of people may need a stand-alone e-mail management program to help them. Here are some suggestions:

PC iMail	**www.prosoftapps.com**
NetMailer	**www.netmailer.com**
Campaign	**www.arialsoftware.com**
Emailrobot	**www.emailrobot.com**

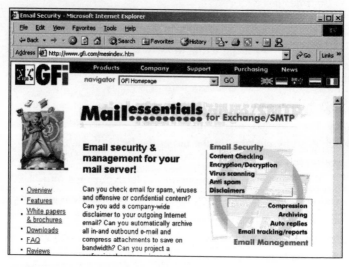

E-mail management software is becoming increasingly important for businesses as the volume of messages increases.

Dealing with 'spam'

One of the disadvantages of making something easy to do is that it opens up new opportunities for life's undesirables. Included in this category are people who buy up huge lists of e-mail addresses then bombard everyone on that list with unsolicited mail usually of dubious value. One CD-ROM can contain some 60 million addresses.

This electronic equivalent of junk mail is known as **spam** and it has become a big issue on the net. It ranges from harmless attempts at mass-marketing to downright fraud, such as pyramid schemes promising quick riches and advance fee scams. Some surveys estimate that spam is costing business billions and billions of dollars a year in time wasted dealing with it.

There's been such a fuss made, particularly in the US where privacy is a major issue, it has led to increased pressure on web service providers not to sell or divulge e-mail addresses, or at least to give surfers the chance to say whether or not they agree to the selling on of their personal details, including e-mail addresses. Some spammers have been banned by ISPs for jamming up their servers with bulk mailings and leading to a flurry of complaints from members. There are moves to outlaw the practice altogether.

> **TIP**
>
> *If you receive spam (unsolicited e-mail), don't reply to it — if you do, the spammer knows your e-mail address is in use.*

At least with unsolicited e-mail it's easy to delete it quickly. But it can become annoying. There are some steps you can take to filter it out using your mailer settings, but these ruses are easy to avoid by clever spammers. The golden rule it never to reply to it. Until you do, spammers don't really know that you exist, given the high turnover in e-mail addresses. No response at all is the best response. Never let a spammer know where you live and if you are receiving indigestible amounts of spam, complain to your ISP – they might be able to block incoming mail from known spamming sites.

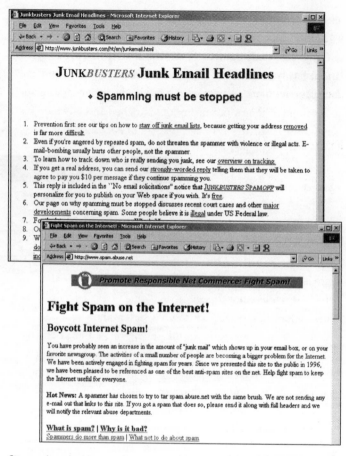

Sites such as Junkbusters (top, www.junkbusters.com) and Fight Spam on the Internet (bottom, http://spam.abuse.net) give advice and programs to help filter out spam.

Having said that – and I know this is tempting fate – I've written an internet column for *The Sunday Times* for over two years, leaving my e-mail address at the end of it each week, and I've not been troubled by spam at all. So it's not an issue to worry about too much, just don't go giving your e-mail address away too freely.

For a plethora of information on spam and how to deal with, including links to all the programs that can help filter out spam, go to **Spamfree** (www.spamfree.org), **Junkbusters** (www.junkbusters.com), or **Fight Spam on the Internet** (http://spam.abuse.net).

Free e-mail everywhere

These days it seems that everyone is trying to offer us free e-mail addresses, from portal sites to search engines. They are mostly web-based services, called **webmail**, that you can access from any computer without the need to mess about with mail server address settings in the mailer program.

In fact you don't need a mailer program at all, you can access your mail from the web browser. All your messages are stored on the web, so it's ideal for people on the move. It also means that if you change your ISP you don't have the aggravation of having to tell all your contacts your new e-mail address – your webmail is independent of any ISP.

But there are downsides. For example, if you archive a lot of e-mail for reading offline, webmail isn't really suitable, unless you're prepared to stay online for long periods of time.

And with most of these free services you have to put up with online adverts cluttering up your screen. Another potential problem is that you are at the mercy of the web. At busy times of the day it can take a long time to access your account and access your messages. These services are truly international so they can have millions of people trying to access their accounts at the same time. With a local ISP the network loop will be shorter, and so you should be less affected by delays.

The most successful of the webmail services is **Hotmail** (www.hotmail.com), which is owned by Microsoft – you can

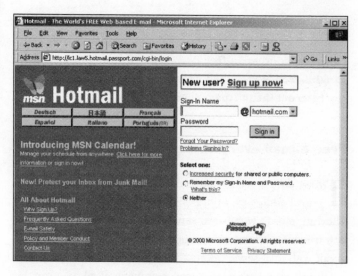

The most successful free webmail service is Hotmail (www.hotmail.com), owned by Microsoft.

set up a Hotmail account within the latest version of Internet Explorer. It's so popular that getting the user name you want is difficult because millions of people have been there before you – it requires great imagination!

But it is handy having a back-up e-mail account if your ISP's mail service crashes, as it will from time to time. It is also useful if you want to be anonymous while on the web, because you can enter whatever details you like, including a false name. Even this isn't a complete cover though, because it is still possible to identify you by tracing your computer's **IP address**.

TIPS

The advantage of a web-based e-mail account is that you can send and receive mail using any computer with internet access.

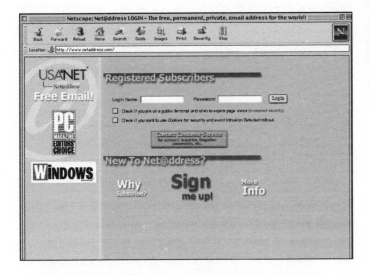

Other free webmail services include Net@ddress (above,
www.netaddress.com) and Yahoo!Mail (below, http://mail.yahoo.com).

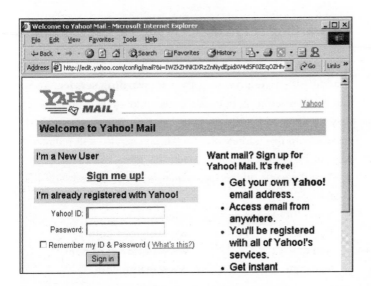

Here's a list of some other free webmail services:

Yahoo!Mail	**http://mail.yahoo.com**
Lycos Mailcity	**www.mailcity.com**
MailExcite	**www.mailexcite.com**
AltaVista	**http://mail.altavista.com**
GoPlay	**www.goplay.com**
Net@ddress	**www.netaddress.com**
SuperNews	**www.supernews.com**
CoolList	**www.coollist.com**
Rocket Mail	**www.rocketmail.com**

For a more complete list try an excellent site called **The Free E-mail Providers Guide** (www.fepg.net) which has tons of information on over 1,000 free e-mail providers around the world. There's a country-by-country breakdown and tables giving the main features of each service. Another useful all-round site is **Email Addresses** (www.emailaddresses.com).

Forwarding services

If you're happy with your ISP but you don't like the e-mail address they've given you, try a **forwarding service**. You can get yourself a more exotic e-mail address and quote that as your main address to all your contacts. When people send messages to this address the forwarding service will redirect it to your ISP e-mail account. Again, this gives you the freedom to keep the same address no matter how many times you change ISP.

Some forwarding services include:

Bigfoot	**www.bigfoot.com**
NetForward	**www.netforward.com**
Mail.com	**www.iname.com**

For how to keep your e-mail private and secure, *see* **Safe Surfing,** *page 159.*

Forwarding services Mail.com (top, www.iname.com) and Bigfoot (bottom, www.bigfoot.com).

Chapter 7

Making the Most of the Web

Introduction

The net is a fabulous resource and a lot of fun. The experiences it offers are becoming richer by the day as new technologies come on to the market and increase its versatility still further. In this chapter, we tell you how to make the best use of what the net can offer, from news to music, chat discussion groups to online shopping.

Unlike other net guides, this book isn't just a list of wacky websites that will need to be updated immediately after it's gone to print. The URLs chosen are simply a selection of the best ones around. If you're looking for a more comprehensive list of sites just bookmark a site like **UK Directory** (**www.ukdirectory.co.uk**). It lists thousands of categorized websites that are constantly updated and expanded.

If you're after the weird and arcane you'll find it pretty quickly on the web anyway – there's enough of it around. And if you want to see what all these other millions of surfers do with their time, go to a site like **Hot 100 Websites** (**www.hot100.com**), which lists the most frequently visited websites each week.

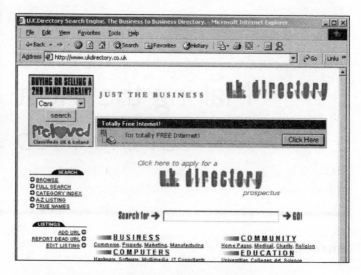

UK Directory (www.ukdirectory.co.uk) lists thousands of categorized websites that are constantly updated and expanded.

So fire up your browser and make the net work for you. There's no need to be intimidated – it's just a pretentious telephone after all. And once you've mastered the basics you can then move on to one of our niche guides for a more in-depth look at particular subjects, such as money or shopping.

News and information resources

The net was invented and supported largely by academics wanting to exchange and share information. And this is still one of the net's greatest achievements. The sheer volume of information out there is awe-inspiring. Most newspapers are now online, as well as having net-only services, such as personal finance or technology-related sites. There are encyclopaedias, libraries, dictionaries, museums and government departments online. Information, from weather forecasts to share prices, has never been so readily accessible thanks to this electronic medium.

Organizations have realized that they can save themselves a lot of bother by putting as much information on the web as possible, where people can help themselves without jamming up switchboards and wasting staff time.

The net can tell you what's happening this second via the many newswire services now available, as well as providing a valuable archive resource when you're researching a topic. What's more, the net allows information services to be much more interactive and users much more selective in what they receive.

For example, you can subscribe to any number of specialist online magazines and newspapers and have news delivered to your e-mail in-box. These services are often called **mailing lists**. You can pre-select the type of stories you're interested in, too. In fact, it has never been easier to

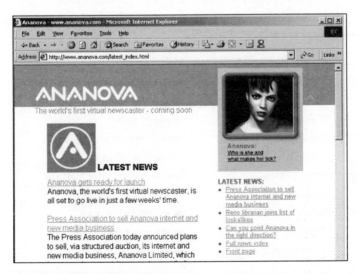

Is this the future of news online? Cyber newsreader Ananova from the Press Association.

keep bang up-to-date with events in whatever field you're interested in.

Just make sure that if you do subscribe to mailing lists you save the e-mail that tells you how to unsubscribe,

otherwise you may find your in-box filling up even after several attempts at stopping your subscription.

Your ISP's home page is a good place to start looking for information resources. It's in their interests to make their pages and services as comprehensive and useful as possible. The latest versions of web browsers also contain lots of pre-selected links to radio, news and entertainment resources. These are sometimes called **channels**, although in my view, this attempt to make the web analogous to television confuses rather than helps.

Anyway, there are hundreds of news and research resources on the web, but here are some prime candidates for your bookmarks folder:

National newspapers

The Sunday Times	**www.sunday-times.co.uk**
The Times	**www.the-times.co.uk**
Financial Times	**www.ft.com**
Telegraph	**www.telegraph.co.uk**
Guardian	**www.guardian.co.uk**
Independent	**www.independent.co.uk**
Daily Mail	**www.dailymail.co.uk**
Mirror	**www.mirror.co.uk**
Express	**www.express.co.uk**
Evening Standard	**www.thisislondon.co.uk**

Live news

BBC	**www.bbc.co.uk**
ITN	**www.itn.co.uk**
PA Newswire	**www.pa.press.net**
Reuters	**www.reuters.com**
CNN	**www.cnn.com**

News aggregators

BBC Monitoring (world news)

	www.monitor.bbc.co.uk
NewsNow	**www.newsnow.co.uk**
Newswatch-UK	**www.newswatch.co.uk**
NewsHub	**www.newshub.com**

Technology News & Round-ups

Wired	**www.wired.com**
Internet.com	**www.internet.com**
NUA Surveys	**www.nua.ie**

For the news on Sunday plus reviews for the rest of the week, go to the
Sunday Times website (www.sunday-times.co.uk).

Different ways of catching up with the news: Newswatch-UK (above, www.newswatch.co.uk) is a news aggregator, while the more specialized Wired (below, www.wired.com) gives technology news and round-ups.

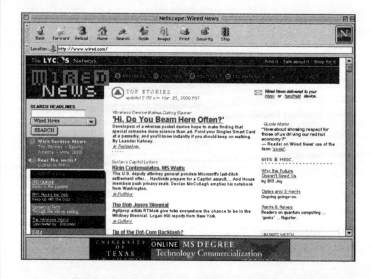

Research resources

Encyclopaedia Britannica **www.eb.com**
www.britannica.co.uk

CCTA Government
Information Service **www.open.gov.uk**
Electronic Yellow Pages **www.eyp.co.uk**

Usenet: news and discussion groups

There is another resource on the net that no other medium can provide: other users. The other major part of the net besides the web is **Usenet**, the collective name for all the **discussion groups**, **bulletin boards** and **newsgroups**. When you download 'news' on the net, it doesn't mean news in the conventional sense, it's the collective name for newsgroup messages. By the same token, messages are often called 'articles' as well, just to confuse things further.

People can read and send messages to each other via these public forums, categorized according to subject. There are over 60,000 separately listed areas of interest ranging from the erudite to the barmy. But they are open to the world and there is a wealth of opinion and experience waiting to be tapped if you know where and how to look. It's not just all text messages either, people can attach picture, video and sound files to their messages, too. You'll need all the right plug-in software to handle multimedia files though (*see* **Downloading Software**, *page 75*).

When you're online to newsgroups you're not on the web, you're just talking directly to your ISP's news server. But you can also access all these newsgroups through the web, by going to sites such as **Deja.com** (**www.deja.com**), the best-known of all the newsgroup websites. This at least gives

You can access newsgroups through websites such as Deja.com (www.deja.com).

you more flexibility in that you do have access to the web as well, but you don't have as much control to organize your news messages as you do within a dedicated newsreader.

Of course, the volume of information can be overwhelming, so you need to marshal these resources carefully and be selective, otherwise you find yourself spending all day trying to read everything published, not getting any work done, and running up huge telephone bills. The net is very good at giving you the illusion of working when you're really just wasting time!

Usenet newsgroups are like public forums where you can post messages and read all the other messages. Anyone opening the newsgroup folder can read your message and reply publicly or to you directly by e-mail. Your messages could be read by anyone.

Setting up your system to access newsgroups

First you need a **newsreader** program. Luckily, the latest versions of Internet Explorer and Netscape Communicator include newsreaders that should be fine for most people's needs. Most ISPs have a dedicated news server, so you have to tell your newsreader how to recognize the server by entering its address. You may have done this already when you first signed up with your ISP. If you didn't, ask your ISP for help.

Once you've entered the name of the news server – usually 'news.ISPname.net' or something similar – you go online to download all the newsgroups that your ISP has stored on the server. You can then search for a subject you're interested in and 'subscribe' to that newsgroup, if one exists.

The discussion groups are categorized according to their content. The abbreviation at the start of the group name, tells you the main subject area. For example, **sci.** stands for 'science' and **alt.** stands for 'alternative', meaning off-the-wall. As you move right through the newsgroup name you go down through subdirectories to more specific subjects within that category. So the longer the address, the more specific the subject. The newsreader search facilities are pretty good so you don't need to memorize newsgroup addresses. And once you've subscribed to groups they appear listed in your newsreader anyway. Just click on the one you want and it will take you there.

> **TIP**
>
> *Usenet newsgroups are like public forums where you can post messages and read all the other messages. Anyone opening the newsgroup folder can read your message and reply publicly or to you directly by e-mail.*

Reading news offline

Your phone bill can mount alarmingly if you stay online while browsing through messages. You can tell your newsreader to download all the messages at once when you go online so that you can log off and read them offline at your leisure without worrying about the bill.

In the Outlook newsreader choose 'Work offline' from the 'File' menu before you go online. Choose one of the groups you've subscribed to, then right-click on it and choose 'Properties'. If you click on the 'Synchronize' tab you're given several options, depending on how many of the messages you want to download. Go online, click on the group, then choose 'Synchronize' from the 'Tools' menu. When you're offline again you'll be able to read all the messages you downloaded by selecting the 'Work offline' mode again. It's one of those things that's easier to do than describe!

In Messenger go online and right click on a newsgroup you've subscribed to. Select 'Newsgroup properties', then 'Download Settings' and set your preferences. Click on 'Download Now' and messages will be downloaded into your reader. Once offline, choose offline mode from the file menu and browse away.

Posting a message

You can contribute to discussions or start a new one – known as starting a new **thread** – and also reply privately by e-mail. For tips on how to disguise your e-mail address when contributing to discussions, *see* **Safe Surfing**, *page 159*. Give your message a relevant subject header to make it easy to identify. To start a new message just click on the 'New Message' tabs in your newsreader menu bar. You can send messages to several groups at once by putting them in the 'CC box' as with e-mail. Once you've sent a message you can also delete it from the newsgroup by right clicking on it and choosing 'Cancel Message', but beware – the posting might still make it onto a few servers, and get read and replied to before it gets cancelled.

Following the rules

Avid Usenet users are pretty intolerant of 'newbies' – new users – coming in and flouting the conventions. So if you want to get along with everyone, make sure you read the Frequently Asked Questions (FAQ) file in the newsgroup if there is one, and make sure you have hit on the right group for the subject you're interested in. You don't want to waste people's time with irrelevant messages. There are a few no-nos when sending messages, such as not typing in capitals – known as shouting – and not being needlessly impolite.

Chat

So-called chat rooms are places where you can engage in almost instantaneous conversations with people by typing in messages and reading their replies. ISPs will often host web chats for their users, inviting guests to take part. AOL and CompuServe have their own chat forums

WARNING

Usenet users can be pretty intolerant of newcomers, so make sure you read the Frequently Asked Questions file in the newsgroup and follow the group's conventions as far as possible.

which are safe, controlled environments for you to start in if you're uneasy about the whole concept. Increasingly companies are using chat rooms as a way of providing customer service online – it's often cheaper to communicate this way than by telephone, though seldom quicker.

Chat rooms are usually organized along subject lines. Go into one room and they'll be talking about one topic, go into another and the subject may be entirely different. Lots of people can be logged on at once, often leading to lively conversations. Eavesdropping on some of these conversations sometimes makes you wonder what people do all day, and who pays the telephone bills! Often it's not the chatters.

The most common chat software program is called **Internet Relay Chat (IRC)**. You can download it from **mIRC** (**www.mirc.com**) if you use a PC, or **IRCLE** (**www.ircle.com**) if you use a Mac. There is lots of help on these sites on how to configure the software and how to find chat servers. The program's own 'Help' file is also useful. You have to give yourself a nickname and also decide whether you want to use your real name and e-mail address at all.

When you enter a chat room your nickname is automatically logged so that everyone else in the room can see you've arrived. Take time to learn how IRC works or else you might end up annoying everyone by entering the wrong commands or earn their general contempt. Again the IRC 'Help' file will tell you most of what you need to know. Generally the system works on the basis of a forward slash

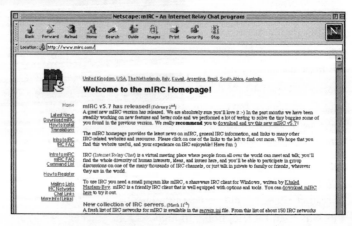

IRC is one of the most well-known chat software programs.

*If you want to know everything that's happening in Chatland try a
site like TalkCity.*

signifying that a command follows, such as /JOIN or /EXIT.
You can have private conversations within the room, set up
your own chat channel and control who you allow in. There
are numerous chat conventions and shorthands that can be
quite intimidating at first, but you'll soon get to know them if
you observe others closely, and they're not compulsory.

Rather than describing how it all works here, just
download the software, read all the 'Help' files, go online and
read some more FAQs on the IRC sites mentioned above,
then try it out.

If you want to find out what's happening in chatland, visit
some of these sites:

TalkCity	**www.talkcity.com**
Yack	**www.yack.com**
The Globe	**www.theglobe.com**
Liszt	**www.liszt.com/chat**

Webchat

There are an increasing number of places you can enter chat rooms without having to have special software. The rules are generally simpler and it can be less fuss all round. Search engines and portals in particular are adding chat facilities to their range of services.

Private chat

Another popular way to chat online, but more privately, is to use a program like **ICQ** (**www.mirabilis.com**) or **AOL Instant Messenger** (included with the Netscape Communicator browser). When you're online these programs will detect whether any of your friends are online, too, so that you can send a message to them instantly and say hello. You need to set up a 'Buddy List' so that the program can recognize who is your friend and who isn't.

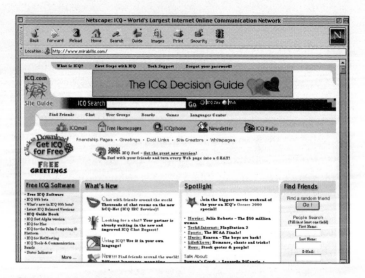

Programs such as ICQ (www.mirabilis.com) can detect whether any of your friends are online so that you can send a message to them instantly.

Shopping online

Shopping online is finally taking off after many false starts. Security fears surrounding using credit cards online are gradually receding as people become more aware of encryption and how to use their cards safely online (*See* **Safe Surfing**, *page 159*, for how to shop safely).

These days there are very few things you can't buy over the net, from groceries to garters, CDs to cigars. It's all there if you look hard enough. Shopping online can be a convenient, simple, safe, and cheaper way of shopping. You don't need to suffer the indignities of public transport, queues, elbowing crowds and unhelpful sales staff any more. There are no such nuisances on the net. There are sites that will even arrange for your purchases to be gift-wrapped. And the net never closes. Nor does it recognize any geographical boundaries, literally opening up a world of opportunities. Only the tax authorities bother with boundaries, but that's another matter.

Some research companies reckon that in the UK alone we'll spend up to £10 billion online by the end of the year 2000, more than 20 times the amount we spent online in 1998. Most high street retailers are scrambling to get their stores online, with mixed results. Designing yourself a website that's fun and easy to use is hard enough, putting in place the delivery mechanisms you need to fulfil all the orders accurately and speedily is even harder.

It has to be said that many retailers have rushed online without getting the nitty-gritty sorted out first, leading to a lot of disillusioned shoppers. Online shops can run out of stock just as easily as ordinary shops. But they are now beginning to realize that online shopping isn't just an extension of their existing businesses, it's an entirely new business, and needs to be funded as such.

Interactive digital television, offered by the likes of Open and ONDigital brings online shopping into the living room. The theory is that television watchers, of which there are millions more than net surfers, will be more comfortable shopping this way. It's all about convenience and ease of use. Surveys show that people are even prepared to pay a premium for the convenience of shopping online. You certainly shouldn't assume that everything is cheaper on the web, but there are some fantastic bargains around.

FACT

Some research companies reckon we'll spend up to £10 billion online by the end of the year 2000, which is more than 20 times the amount we spent online in 1998.

For example, you can buy CDs and videos from US online stores, such as **CDNow** (www.cdnow.com) and **Amazon.com** (www.amazon.com), for the dollar equivalent of sterling prices, even after taking delivery and tax costs into account. And you can save hundreds of pounds on electronic goods, such as digital cameras and computers by shopping online at US sites and importing to the UK, although you sometimes find that retailers over there won't ship internationally.

So where do you start? There are plenty of 'shopping-mall' sites that aggregate hundreds and even thousands of online retailers, often reviewing and rating them for ease of use and quality of service. Try these for starters:

I Want To Shop	www.iwanttoshop.co.uk
2020Shops	www.2020shops.com
ShopSmart	www.shopsmart.co.uk
Yahoo! Shopping	http://uk.shopping.yahoo.com
IMRG Shops Directory	www.imrg.sotn.com

And *see* **Navigating the Net,** *page 45* for a list of shopping agents that can help you compare prices across a number of

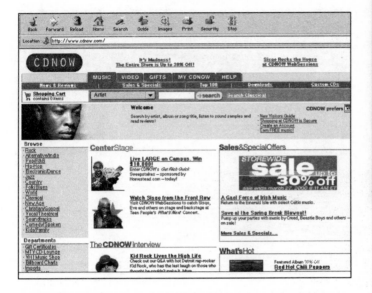

You can often buy CDs more cheaply from the States at sites like CDNow.

retailers. These price comparison agents are applying considerable downward pressure on prices because there's nowhere to hide on the net. If you're a retailer and your prices aren't competitive, a shopping agent will spotlight it instantly. More power to the consumer!

Shopping abroad

Before you get carried away flashing the plastic online, bear in mind that there are all sorts of taxes and charges to consider when buying abroad. There may be local sales taxes (in California for example), then there's the delivery charge. This will usually be a tiered rate depending on whether you want standard or express delivery. And once the goods reach these shores Her Majesty's Customs & Excise pounce and may levy import duty and VAT. The rules are very complicated, so before you order, check out the potential charges at the **Customs & Excise** website: **www.hmce.gov.uk.**

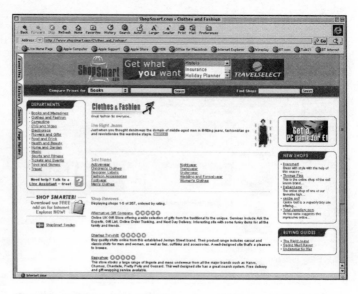

*'Shopping mall' sites such as ShopSmart (www.shopsmart.com) aggregate
hundreds of online retailers.*

Books are zero-rated for import duty and VAT, so it can be
cheaper buying from US retailers, but cigars always attract
excise duty. If you're not careful the final bill for your hand-
rolled Cubans could give you a heart attack.

Shopping abroad also has its drawbacks. Supposing the
product you bought was faulty? Warranties often only extend
to the country the product
was sold in. Even if it did
extend to the UK, sending
it back for repair is
obviously going to be
time-consuming and
expensive. Similarly for
orders that have just been
executed wrongly. Also, credit card issuers dispute
whether the protection afforded by the Consumer Credit Act

WARNING

*The price of goods you see on the net
may not be the price you would pay. You
may need to take into account sales tax,
delivery charges, import tax, VAT...*

extends to foreign credit card purchases (*see* **Safe Surfing,** *page 159*).

Online auctions

One of the fastest-growing ways to shop online is through auctions. The number of sites is mushrooming to cater for millions of bargain-hunters and collectors, buying, selling, and swapping anything from bric-a-brac to high-quality antiques. General retailers, holiday companies and airlines are realizing that auctions are an excellent way to get rid of surplus stock or products they can't sell elsewhere. Selling something at a knock-down price is still better than not selling it at all, they reason. And individuals are realizing that auctions are a fantastic way to gain access to a wider audience and a wider choice of things to buy.

There are auction sites specializing in fine art and others content to be an extension of the classified ads system. Usually online auctions take days or even weeks, but some sites let you bid live via your PC at real in-house auctions so that you can participate in the sweaty-palmed excitement even if you can't make it in person.

Most online auctions work in much the same way. First you have to register, which means filling in an online questionnaire, giving personal contact details, and maybe a credit card number. You also have to agree to the site's terms and conditions. It's tempting to skip these as they are rather dull, but you should find out about the site's charges and its payment and delivery policies first before bidding.

Here are some of the leading auction sites, although there are hundreds more:

eBay	**www.ebay.com**
	or **www.ebay.co.uk**
Amazon.com	**www.amazon.com**
	or **www.amazon.co.uk**

Priceline	**www.priceline.com**
QXL	**www.qxl.com**
eBid	**www.ebid.co.uk**
IbidLive	**www.ibidlive.com**
iCollector	**www.icollector.com**
Loot	**www.loot.com**
The Auction Channel	**www.theauctionchannel.com**
Yahoo!	**http://uk.auctions.yahoo. com/uk**
Sothebys.com	**www.sothebys.com**
Sothebys.amazon.com	**www.sothebys.amazon.com**
FiredUp.com	**www.fired.com**

If you register with a lot of auction sites, a service like
Bidstream (**www.bidstream.com**) could help. It's a
dedicated auction search engine that can help you sift
through all the available auctions and simultaneously monitor
ongoing bids you've made on different sites.

Telephoning across the web

You can use your computer to telephone people, too. The
difference is that your voice is sent across the telephone
network in a different format – the same way that internet
data is sent. This format is called **Internet Protocol** (IP). Net
telephony products are sometimes called IP telephony, Voice
over the Internet (VOI) or Voice over IP (VOIP).

Net telephony does not yet offer the same quality of
telephone service as direct telephone connections, but things
are improving fast. And there are more and more net
telephony applications around, such as CoolTalk and
Microsoft's NetMeeting, that often come incorporated in the
latest versions of web browsers.

Whether it's fine art or fitness equipment you're after, there's an auction site to suit you.

There are two categories of IP phone call – those that use the net for the entire length of the journey, and those that go part of the way on the net before joining the normal telephone system at some point. The great thing about pure net calls is that you can phone friends or business partners in far-flung corners of the world very cheaply, because you're still linking up to your ISP at local call rates.

The only drawback, and the reason why it will take some time before net telephony takes off in a big way, is that pure net calls require both your computer and the computer you're 'ringing' to have special software and equipment to enable them to send and receive voice calls. You need a good sound card in your computer plus a high-quality microphone and speakers. Ideally, a headset – where the earphones and microphone are in one piece – will ensure better quality all round.

The standard PC sound card usually allows only talking or listening (half-duplex), not both simultaneously. This can make for slightly stilted conversations. To replicate the ordinary telephone experience you need a 'full-duplex' card that you then have to install in your computer. Some advanced sound cards already have IP telephony software built in. Once you've got the right hardware you can make calls to other computers through net telephone systems such as **ICQ** (**www.icq.com**). **VocalTec Internet Telephone** (**www.vocaltec.com**) is another well-known service.

Another difference from conventional telephone calls is that you usually have to schedule the call so that you can both arrange to be online at the same time. If not, you simply won't be able to get through. In the US this isn't so much of a problem because they generally get free local calls allowing them to stay online permanently. In the UK, where free calls have yet to become standard, net telephony takes a little more planning.

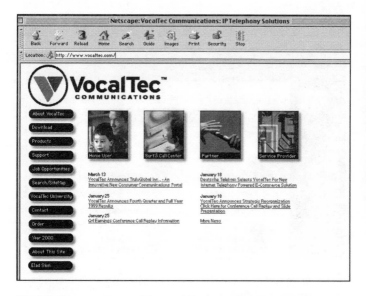

Vocal Tec (www.vocaltec.com) Internet Telephone allows you to make telephone calls to other computers.

Also, the quality of your connection will only be as good as the state of the net at that time. When the network is busy you may experience delay in hearing the other person, or parts of their words may drop out. This is because IP involves your voice being segmented into little packets of data that are then sent individually across the network by the shortest available route. This inevitably means that some packets may get log-jammed on their journey, leading to an imperfect voice quality.

WARNING

If you use net telephony, the quality of your connection will only be as good as the state of the net at that time. When the network is busy you may experience delay in hearing the other person, or parts of their words may drop out.

You can make telephone calls from your computer using services such as Delta Three (www.deltathree.com).

Net-to-phone telephony

Luckily technology does allow you to make calls from your
PC to ordinary phones via so-called 'hop-off' services. These
provide gateways between the net and local phone systems
in other countries. Such companies do charge for their
services, but you're still guaranteed to make savings on
international calls.

There are a number of companies that specialize in
providing this type of service. Two of the most well-known
companies in this area are IDT, which makes **Net2Phone**
(**www.net2phone.com**) and **Delta Three**
(**www.deltathree.com**).

If it's just cost you're concerned about, rather than the
convenience of using your PC to make calls, you can even
use your ordinary phone to ring people across the net using
a service such as **Inter-Fone** (**www.inter-fone.com**). Again,
the drawback is that both phones need to be connected to
the PCs that contain Inter-Fone's own sound cards.

Downloading digital music

The net is transforming the music industry almost overnight thanks to the ease with which we can download music files. We can sample songs before we buy the CD, and buy songs individually, too.

MP3, a digital format for music, is rapidly becoming the standard for the industry. You can now download MP3 files, store them on your computer or transfer them on to portable digital music players, such as the Diamond Rio. There are no moving parts because it's all digital and you get near CD-quality. For tons of information and links to music sites, just go to **MP3.com** (**www.mp3.com**).

First you need a special software plug-in to listen to music. The latest web browsers have them already built in. For example, Microsoft Internet Explorer includes **Windows Media Player** (**www.microsoft.com/mediaplayer**) and Netscape Communicator has **Winamp** (**www.winamp.com**). These enable you to listen to live radio stations – called **streaming audio** – as well as downloaded music files.

There are alternatives, the most popular being

> ### WARNING
> *Music files are very big. Downloading a three-minute song on a 56kbps modem takes around eight minutes.*

RealPlayer (**www.real.com**), suitable for both PC and Mac users. Apple's own media player is called **QuickTime** (**www.apple.com/quicktime**) and **MacAmp** (**www.macamp.com**) is also tailored for Mac users.

Digital music gives much more choice and flexibility to consumers. There are now plenty of music retailers who will make you a CD to order, containing tracks that you select from a menu. You can usually browse by music genre, artist or song title and simply click on the songs you want, paying

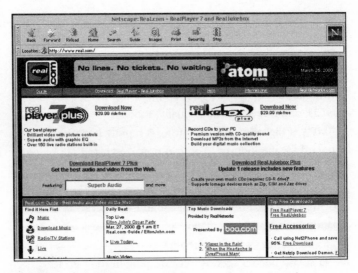

*Enter the world of digitally downloaded music with programs such as
RealPlayer and RealJukebox.*

You can even create your own tailor-made CDs.

per track. Each CD can take up to 74 minutes of music. You then place your credit card order over a secure, encrypted connection and your custom CD will be sent to you.

For a taste of what's on offer try some of the following sites:

MusicMaker	**www.musicmaker.com**
CustomDisc	**www.customdisc.com**
Razorcuts	**www.razorcuts.co.uk**

The one major warning to give about music files is that they are very big. Downloading a three-minute song on a 56kbps modem takes around eight minutes, or 15 to 20 minutes with a 28.8kbps modem. If you want to use a **ripper** program, such as RealJukebox, that can transfer music from your CDs on to your hard drive, you'll need lots of disk space. A typical pop song takes up 2MB to 4MB of space. Just bear this in mind before you decide to put your whole collection on your PC for listening while you work, or before you download entire albums during busy times of the day.

The web goes mobile

You can also access the net from your mobile phone these days, using it to surf specially created web pages designed for small screens. You can send and receive e-mail and make use of a range of interactive services, from share price alerts to the latest football scores. The standard for mobile-friendly web pages is called **Wireless Application Protocol (WAP)**.

Before too long WAP-enabled mobile phones could take over from PCs as the main way people access the net around the world, so web service providers have been scrambling to develop WAP-friendly pages and services. For more

Internet-enabled mobile phones and personal organisers could be the main way we access the net in future.

information on WAP and how to buy a WAP phone try the
mobile phone operators first:

Vodafone	**www.vodafone.co.uk**
BT Cellnet	**www.btcellnet.co.uk**
One-2-One	**www.one2one.co.uk**
Orange	**www.orange.co.uk**
Virgin Mobile	**www.virgin.com/mobile**

Vodafone, the UK's largest mobile operator with around seven
million subscribers, could find itself becoming the UK's largest
ISP, putting the likes of Freeserve in the shade. It has already
launched **Vodafone Interactive** (**www.vodafone.net**), which
allows customers to specify via the website what information
services they want sent to their mobiles.

Virgin Mobile is taking a slightly different approach,
equipping its phones with special chip cards inside. These
incorporate a mini web browser capable of reading WAP
websites. The advantage of this is that you don't have to buy
an expensive WAP phone to receive WAP services.

The heartening thing about WAP is that for once Europe has
stolen a march on the US. European mobile companies have
done a much better job than their US counterparts in agreeing
technological standards. Mobile phone penetration is also much
higher in Europe than in the US. And Nokia, Siemens and
Ericsson are now co-operating on a third generation of mobile
phone capable of downloading net content more than fifty
times faster than current mobiles can manage.

We'll be able to watch live news broadcasts and film
trailers on our handsets, shop securely, and access much
more visually-rich web pages. Voice-activated services will
mitigate the obvious disadvantages of a mobile phone's small
keypad and screen, giving us almost complete freedom to
roam whilst having an impressive array of communication
and transactional resources at our fingertips.

Chapter 8

Safe Surfing

Introduction

When we talk about security on the net, there are several issues involved. Firstly, there is the issue of privacy – preventing our e-mail from being read by others, and our personal details abused by websites. Then there's the problem of how to protect our children from exposure to unsuitable web content.

Thirdly, there's the issue of secure online shopping – using our credit and debit cards online safely without fear of those details being stolen. Fourthly, there's a big issue surrounding authenticity – websites and surfers proving that they are who they say they are and that they will deliver what they promise to deliver. And lastly, there are those dreaded viruses to worry about – malicious or unintentionally damaging programs – threatening to invade our hard drives and reduce them to so much digital rubble.

These are all serious issues. But they are being tackled and largely overcome by technology and public awareness. Yet the public perception is that the net is besieged by millions of fiendishly clever computer hackers breaking into computers everywhere, stealing our money, reading our e-mail, corrupting our hard drives, and compromising the defence of our nations on a daily basis. This just isn't true.

Yes, there have been a number of high-profile cases of credit card numbers being stolen from online retailer databases and published on the web. Anyone using those numbers can in theory go on spending sprees until the cards are cancelled. And yes, hackers do consider it a matter of professional pride to try to crack the best security systems that software companies can throw at them. But their activities are, in the main, not malicious. In fact many argue that they are acting in the public interest by pointing out these security flaws. After all, software companies make great claims about the impenetrability of their security software when they flog it to companies. So shouldn't those businesses be made aware if they are being sold a pup?

It's a moot point, and there's no getting away from the fact that breaking into other people's computer systems is still against the law. But the net result – if you'll pardon the pun – is that these software companies are forced to work harder to improve their products, and we, the public, benefit in the long run because we have a safer environment in which to surf.

TIPS

Many virus warnings are hoaxes. For a comprehensive list of myths, hoaxes and urban legends have a look at www.kumite.com/myths.

It seems that not a day goes by without yet another rumour of a super-virus sweeping the net, each more potentially damaging than the last. Many of these rumours turn out to be hoaxes. Again, I don't want to belittle the security issue. Viruses do exist and they can cause considerable damage. But a good anti-virus software package, regularly updated, can protect you adequately. They're not that expensive, and are easy to download and operate.

The problem is that the media in general like nothing more than a good net scare story. The net is still new so it's exciting to write about, yet many are still largely ignorant about it. And as we generally fear what we don't know, it becomes a vicious circle of misinformation leading to a disproportionate level of suspicion and concern.

The bottom line is that there are ways to deal with these security issues and protect yourself while you're online with a combination of software and common sense. Nowadays we don't think twice about driving with a seat-belt on – it's a necessary precautionary measure. We generally believe that the benefits of driving outweigh the dangers. There should be no difference surfing the net. Bear in mind also that it is in software companies' interests to overhype security fears to help them sell more of their security software!

Privacy

Anonymous e-mail

It is possible for spammers to glean your e-mail address from web pages, chat rooms, directories and Usenet discussion groups using special software. Some programs making use of the JavaScript language can even obtain your e-mail address when you click on icons in web pages. And you don't have much control over what happens to your sent mail when it arrives at its destination. If it is archived, your address is potentially open to hacking over a long period of time.

One way to stop people getting hold of your address is not to give your address out at all. But often this isn't practical, as lots of sites ask you to register with them first before you can use their services, and this involves giving your name and e-mail address. Another way is to use a webmail account, using an address with a made-up user name (*see* **Eletronic Mail**, *page 103*). Giving a false address is another option, but not much good if you want the website to send you a regular newsletter, for example. You should assume that all mail you send and receive at work is accessible by your employers, so be careful what you say.

When accessing Usenet discussion groups there may be times when you want to remain anonymous yet take part in the discussions. There are a number of services that will help you do this. **Anonymous remailers**, as they're sometimes known, strip all the headers from your message. Headers are all those lines of information at the top of an e-mail that list who sent the message and by what route it arrived with you. Once you've stripped away all these the person receiving the message would not be able to tell who sent the message nor be able to respond. But obviously this kind of e-mail has to be used responsibly and is open to abuse by cyber-stalkers and other online trouble-makers.

Here are a few services to try out:

PrivacyX	**www.privacyx.com**
Anonymizer	**www.anonymizer.com**
Zed-Zed dot Net	**www.zedz.net**
Anonymous.to	**http://anonymous.to**
Junkbusters	**www.junkbusters.com**

TIP

When accessing Usenet discussion groups there may be times when you want to remain anonymous yet take part in the discussions. Anonymous remailers can enable you to do this.

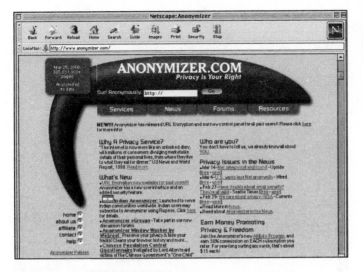

You can go on the net incognito using packages such as Anonymizer (www.anonymizer.com).

Digitally signing and encrypting e-mail

E-mail can be likened to a postcard in that people can potentially intercept and read it while it's en route to its destination. Your ISP could potentially read it while it sits on its servers. For most of us, this wouldn't be a problem even if it did happen, which is unlikely. Not many of us have state secrets to tell after all.

But there are moves to make the **digital signature** and **encryption** of e-mail more widespread. Digital signatures are an attempt to prove that the e-mail really did come from you and no-one tampered with it on the way. Encryption is a way of jumbling up digital data so that it cannot be read by anyone who does not have the key to unscramble the message (*see* **Bluffer's guide to Encryption**, below).

The latest versions of web browsers have encryption facilities already built in. Digital signing and encryption rely on you downloading a **digital certificate** from a 'trusted

If you do need to encrypt your e-mails then companies such as VeriSign
(above, www.verisign.com) and GlobalSign (below, www.globalsign.com) can
help you do so.

third party' – a company such as VeriSign, British Telecom, Thawte, and GlobalSign – that specializes in security products. In the latest version of Outlook you can be taken directly to a Microsoft page that gives links to various certificate providers. Click on 'Tools', 'Options', 'Security', then 'Get a digital ID'. **VeriSign** (**www.verisign.com**) is Microsoft's preferred certificate supplier and you can try out a certificate for a free trial period.

For information on security options in Messenger, click on 'Communicator', 'Tools', 'Security info', then 'Yours' under the 'Certificates' header. Click on 'Get a certificate' and you'll connect to Netscape's site where you'll find that VeriSign is its preferred partner, too.

Here are the addresses for other certificate distributors:

GlobalSign   **www.globalsign.com**
Thawte   **www.thawte.com**
British Telecom   **www.trustwise.com**
(in association with VeriSign)

Once you've downloaded your certificate you can then choose to sign your messages automatically or do so on an ad hoc basis. But for the digital signature to be any use your recipients need to install your certificate on their computers. They can install it next to your entry in the address book. Their mailer programs must be able to support this level of security though.

The problem is that digitally signing an e-mail doesn't necessarily prove it was sent by you. It really just proves that the message originated from your *computer*. Someone else who had access to the computer and who knew your e-mail password could send messages purporting to come from you. This is made all the more feasible by the fact that in your dial-up software you can opt to have your computer remember your password, so that's it's already there in the box when you log on.

So if you really want to protect yourself you should also turn off the 'Save password' option. This is simple to do. When you log on a dial-up box normally appears on your screen as it goes through the process of trying to connect, verify password, and so on. In the latest versions of dial-up software, bundled in with your operating system, there is a box you can tick if you want the system to remember your password. Make sure this box is left unticked. After that, just make sure that you keep your password to yourself.

To encrypt your messages you need to install recipients' certificates on your system as well so that each of you can encrypt and decrypt each other's messages. This is all a bit of a rigmarole and you may well conclude that it's not really worth it, given the largely innocuous nature of your e-mail. It is really much more of a burning issue for businesses at the moment, who could well be sending messages containing commercially sensitive information.

TIP

To stop anyone sending a message from your computer in your name, make sure you have the 'save password' option switched off – and don't go giving out your password!

In the private sphere digital signing and encryption of e-mail is likely to remain a minority sport among technology enthusiasts, at least until certificates become much more widely spread.

Anonymous surfing

When you browse the net, you leave a trail behind you that can be monitored by websites for marketing purposes. Website operators can sometimes find out the IP address of the machine you used to access their site, the type of computer and browser you used, and the last page you were at before you went to their site. They may also be able to determine your e-mail address and real name.

Ask your ISP if it supports **dynamic IP**, which means you are assigned a different IP address every time you log on. This makes if far harder for websites to prove that you are the same person returning to their site. If you stay online for a very long time though, it still may be possible for the site to identify you.

The most common way for websites to log your visit is to send little files called **cookies** to your computer's hard drive when you're online. A cookie is a text file that records your preferences when you use a particular site. Cookies allow the server to store its own identity file on your computer and also for the website to recognize you when you return. This saves you having to log in each time and enables the website to target adverts and services at you according to your preferences.

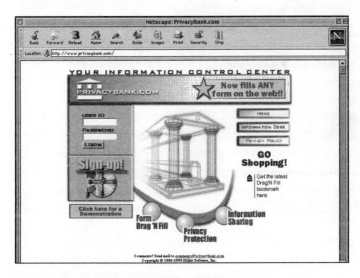

Privacy Bank is a service that interposes itself between you and the websites you are browsing (www.privacybank.com).

The threat from cookies has been overstated – they can be quite useful and most of them don't collect personal information about you, such as your name, phone number or e-mail address. Some web pages won't even load properly unless you set your browser to accept them. But they will record how many times you've been to a site and what you did when you were there.

Cookies are stored in a subdirectory on your hard drive. To have a look at them, go to the 'Cookies' folder in the 'Windows' directory, or the 'Users' folder in the 'Netscape' folder. You can tell your browser how to handle cookies, from accepting them all, to rejecting them all. In Explorer, choose 'Tools', 'Internet Options', 'Security', then the 'Internet zone'. Click on 'Custom level' then scroll down to the cookies section and tick the relevant boxes. In Navigator, click on 'Edit', 'Preferences', 'Advanced' and make your choice.

If you don't like the thought of every website knowing about when you visit them, there are services around that can give you some control over what information you disclose. These services effectively interpose themselves between you and the website. Try some of these:

Lucent Personalized Web Assistant	http://lpwa.com:8000
Privaseek	www.privaseek.com
Privacy Bank	www.privacybank.com
@YourCommand	www.yourcommand.com

TIP

If you want more control over your cookies, allowing those from sites you trust and filtering out the ones you don't, there are several programs around that can help. ZDNet has a lot of shareware versions listed on its site at www.zdnet.co.uk/software, including Cookie Washer, Cookie Crusher, CleanUp!, Cookie Cutter PC and Cookie Pal.

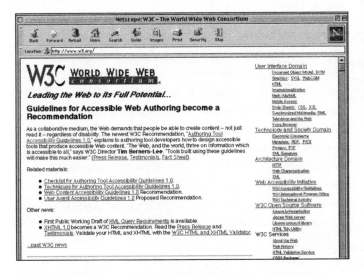

The World Wide Web Consortium (www.w3.org) is in the process of developing a set of privacy standards for the net.

If you want to know more about privacy issues in general have a look at the site of the **World Wide Web Consortium**, known as W3C for short (**www.w3.org**). It's a coalition of leading net companies recognized to be the arbiter of standards on the net. It is in the process of developing a set of privacy standards whereby the net user will be told what the website's privacy policy is before any kind of transaction takes place. It's called the Platform for Privacy Preferences, or P3P for short. Other informative sites include the **Electronic Privacy Information Centre** (http://epic.org) and **Privacy International** (www.privacy.org/pi).

Anti-virus software

Just as every driver should wear a seat-belt, so every surfer should have anti-virus software installed. It's not difficult to do and is essential to protect your computer against invader programs causing a nuisance on your system. Most virus warnings are hoaxes and the threat from viruses is overhyped, but files do become corrupted from time to time even without the hacker's influence. You should be particularly wary of e-mail attachments, especially if they are word processing or spreadsheet documents. The latest anti-virus packages will monitor your e-mail for you and spot attachments containing viruses. They will normally be able to squish or 'clean' the file immediately. ISPs often perform systematic anti-virus checks on your e-mail before you even get it. All in all, you shouldn't get too concerned.

There are three main packages to choose from:

Dr. Solomon's Home Guard **www.drsolomon.com**

McAfee VirusScan (now part of Nework Associates) **www.nai.com**

Norton AntiVirus **www.symantec.com**

Make sure that you have anti-virus software installed. Check out programs such as McAfee VirusScan (www.nai.com) for the latest versions.

When you buy a new computer an anti-virus package will sometimes be included with the software bundle you get. Bear in mind that the program may have been sitting on the hard drive for several months after manufacture and will more than likely be out of date. Visit the software company's website and download the latest file containing all the known viruses, known as the signature file or the virus definition file. Your anti-virus package can detect which file version you have and tell you whether you need to download a new version. Once you've paid the £30 or so for the basic package, updates are usually free.

Filtering web content

For many parents the ease with which unsuitable material can be accessed on the web is frightening. They fear that this box in their homes will corrupt the minds of their children instantly. It is true that pornography and anti-social content is easy to find with a few well-chosen words in a search engine – even with a few innocently chosen words. And there are some bomb-making nutters and unsavoury characters out there. But hey, the net is all about freedom of speech. Freedom is its lifeblood. As with most concerns about the net, there are sensible steps you can take to reduce the risk, and technology is once more coming to our aid.

> **TIP**
>
> *A combination of software and common sense should guard your children against the worst excesses that the net can throw at us.*

Firstly, it is possible to set passwords for your computer so that no-one can get access without you being there. Secondly, you could supervise all online activity and surf the web with your children. Thirdly, both leading web browsers incorporate security features that allow you to filter web content. Using the standard Platform for Internet Content Selection (PICS), they allow you to select the level of content you are prepared to accept in four categories of language, sex, nudity and violence.

In Internet Explorer go to 'Tools', 'Internet Options', then click the 'Security' tab. In Navigator select 'NetWatch' under the 'Help' menu. One problem with these PICS settings is that they can be extremely sensitive. Set them to a moderate security level and you often find that you cannot access the most harmless-seeming of pages. Every time you try to access such a 'banned' page you'll be told you're not authorized to do so without entering the password. This can become extremely tiresome after just a few seconds, so some experimentation is definitely required. I suspect that at this current rather clumsy level most people just disable these ratings after a while. Any system like this will inevitably be subjective. One person's view of offensive is another's idea of a good laugh.

There are stand-alone web filtering packages on the market for around £30 to £40 that can offer a more thorough, less blunt approach. The leading ones are:

Cyber Patrol	**www.cyberpatrol.co.uk**
Net Nanny	**www.netnanny.co.uk**
SurfWatch	**www.surfwatch.com**
Cyber Snoop	**www.pearlsw.com**
CyberSitter	**www.cybersitter.com**

Cyber Patrol will even allow you to specify how long people spend online, and it doesn't just cover web browsing but e-mail and online chat as well. It will also give lists of sites that are suitable for youngsters, as well as screening out those that aren't. Plus it will report any attempts to access forbidden sites.

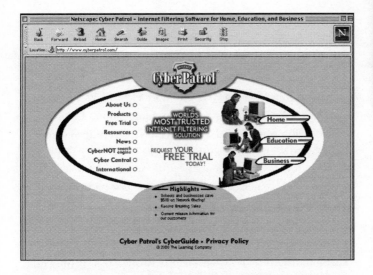

With judicious use of web filtering programs such as Cyber Patrol and SurfWatch, surfing the net can be a safe and rewarding experience for you and your children.

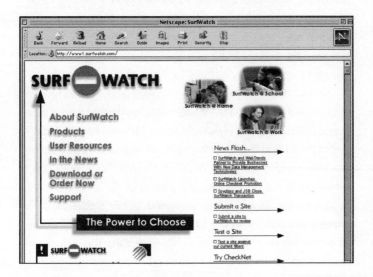

Of course, even the most sophisticated filtering and
monitoring software isn't going to deter the most ingenious
and persevering teenager from finding ways round the
blocks. And there is always the danger that by forbidding
something it just becomes all the more exciting and a greater
challenge. Cod-psychology aside, a combination of software
and common sense should guard your children against the
worst excesses the net can throw at us.

Bluffer's guide to encryption

Encryption is a way of encoding information so that it can
be transferred across an open network, such as the net,
without anyone being able to understand what the message
is. Only the intended recipient has the 'key' to unlock the
code. The data is jumbled up according to a mathematical
formula or algorithm and the way these rules are
implemented depends on a variable cryptographic key – a
string of ones and noughts basically. A 40-bit key has 40 ones
and noughts and a 128-bit key has .... well, you get the
picture.

To find the key you have to guess the precise arrangement
of ones and noughts. This is well-nigh impossible for us mere
mortals but easier for extremely powerful computers which
can crunch billions of numbers per second. Codes have been
cracked, but this shouldn't worry us unduly, because it took a
massive amount of computing power to do it.

A code-cracking machine designed by the Electronic
Frontier Foundation in the US searched more than 88 billion
keys every second for 56 hours before finding the right 56-bit
key to decrypt an encoded message. But to crack a code that
used 128-bit encryption you would need computing power
more than 1,000,000,000,000 times greater than that used to

decrypt a 40-bit message. This is currently impossible and likely to remain so for the foreseeable future.

The key to successful cryptography is not discovering a completely fool-proof code, but making it so difficult that it's not worth anyone's while to try to decode it. As long as the value of the prize is lower than the cost of winning it, encryption should deter criminals. A criminal who can easily get your credit card number from a carbon paper receipt in a shop or restaurant is hardly going to bother linking up 250 computers on the off-chance that he might intercept your number as it whizzes through cyberspace.

There are two main types of encryption: symmetric (secret key) and asymmetric (public key). Symmetric means that the same key is used to encrypt and decrypt the message, and asymmetric means one key is used to encrypt and another to decrypt. Everyone has two keys, a public and a private one. When you want to send an encrypted message you use the recipient's public key, which sits on your computer in the form of a digital certificate. The recipient decodes the message using their private key. It is actually a lot more complicated than that, but you don't really need to know all the gory details.

The digital certificates can also establish the authenticity of the person sending the message and the fact that the message hasn't been tampered with in transit. They are managed and distributed by 'trusted third parties', such as VeriSign and GlobalSign.

Encryption is controversial because the stronger you make it the more governments think it will help criminals to conceal their activities from law enforcement agencies. The US government restricts the transactions that can use 128-bit encryption to online banking and those by financial institutions. For other forms of online transaction we've had to make do with 40-bit or 56-bit encryption.

Secure online shopping

You wouldn't think it, but shopping online is actually safer than handing your credit card over in a restaurant or ordering over the phone. Most online retailers offer secure payment systems now that encrypt your credit details with up to 56-bit encryption. There have been no reported instances of hackers intercepting credit card details while they've been flying across the web.

But survey after survey shows that fear over security is still the main reason why people are reluctant to shop online. This fear is largely unfounded. The best way of dispelling this fear is simply to do it and spread the word.

The main security issue lies with the retailer and what steps it takes to protect your details if it stores them on its systems. There have been several cases of hackers being able to break into online retailers' systems, steal credit details, and publish them on the web. But how much should we worry about this? We are only liable for the first £50 of any loss due to fraud, so long as we haven't been negligent and we've reported any rogue purchases as soon as we discovered them. In a way, it's not our problem – it's the card providers' and the retailers' problem. Of course, it comes back to us in the long run. If the fraud losses mount the cost is inevitably passed back to the consumers somewhere down the line.

TIP

Shopping online is actually safer than handing your credit card over in a restaurant or ordering over the phone.

The standard online payment security system is called **Secure Sockets Layer** (**SSL**), developed by Netscape for its Navigator web browser. Now it is the most widely used system and is supported by all the major browsers.

There are other security systems around, such as the Secure Electronic Transaction (SET) protocol developed by Visa and Mastercard among others. The main advantage of SET is that the retailer can find out if the cardholder is using a valid card without needing to see the credit card details, and the card issuer doesn't know what is being bought, just the price.

Confidence in secure payment systems has grown to such an extent that several credit card providers now offer fraud guarantees with their cards, promising to stand any loss the customer incurs through fraud on the web.

Once strong encryption has become widespread and security of data crossing the net is assured, we'll be able to pay for things in any number of ways. One of the simplest is electronic funds transfer, where cash is taken directly from your bank account, as if paying with a debit card. This way there are no credit card details being transmitted at all. We'll also see 'electronic purses' that store electronic cash being used for smaller online purchases, where credit cards wouldn't be cost effective.

How do I know that an online retailer offers secure payments?

First of all the retailer will probably trumpet the fact very loudly. It wants more than anything for customers to feel that they can buy online in confidence. But don't just take its word – your browser will tell you. You will see a closed padlock at the bottom of your screen indicating that the

> **TIP**
>
> *Look out for a closed padlock or unbroken key symbol on retail sites. This tells you that a site is secure.*

site is secure. You may also see that the web address has changed so that the address begins **https://** rather than the usual **http://**.

Encrypting data takes a lot of computer resources and can slow things down considerably on the web, so what many online retailers do is give you the option of switching between secure and insecure mode. If you're just browsing the store and choosing things to buy, there's no reason for the link to be secure – you're not transmitting any sensitive information. It's when you come to buy using your credit card that you need encryption. So at this point retailers will often have a button giving you the option to switch to secure mode. This is when you should see the web address and the browser security icons change.

How can I trust an online retailer?

Online shoppers do have legitimate concerns about shopping online, but they shouldn't be about using credit and debit cards. The main problem is knowing who you're dealing with. If you've never heard of the site, how do you know that it's genuine, has the goods it says it has and will deliver them? How do you know that they will keep your credit card details secure from theft, internal or external? In short, you don't. You have to satisfy yourself about genuineness and reliability.

In the high street you'd be a little suspicious buying goods at knock-down prices from a bargain-basement shop. We all know what these places look like and we're pretty familiar with the hard sell patter. On the net it's easier to put up a sophisticated and credible shop front. Proving authenticity and integrity is a harder task for shoppers. Some fraudsters have passed themselves off as genuine retailers then gleaned the credit card details from gullible victims and gone on spending sprees using the stolen numbers. It is quite easy to make a website that is virtually identical to a well-known site.

Follow these simple rules to protect yourself:

● Never give your card details over the net except via a secure server.

❷ Never write down or disclose passwords, log-in names or Personal Identification Numbers (PINs).

❸ Stick to well-known, well-regarded websites if possible. Ask friends for recommendations.

❹ If you've never heard of a website and you're unsure about it, look for physical address and telephone contact numbers. Test them to establish that the business really exists. Ask your friends if they've heard of it. If you have any remaining doubts, don't deal with them.

❺ Also check that the web address is exactly right. Fraudsters can sometimes set up virtual copies of well-known brand-name websites. A dot here and a hyphen there can make all the difference. And bear in mind that a .co.uk or .uk ending doesn't necessarily mean that the site is based in the UK.

❻ Look for sites that have been given a 'kitemark' certificate by an accreditation scheme, such as VeriSign, WebTrader from the Consumers' Association, TRUSTe, BBBOnLine, trustUK and JIPDEC. These schemes check out websites for authenticity, security and responsibility in the handling of personal details.

❼ Look for sites that send you an e-mail confirming your order and giving you a unique order number that you can use to track the progress of your purchase. Make sure you keep these e-mails as proof of purchase and for reference if you need to contact the retailer.

❽ Ask what delivery guarantees an online retailer gives and what its returns policy is. If it is a UK site you're covered by normal consumer law, such as the Sale of Goods Act, and entitled to a full refund if goods are faulty or not as advertised. Be extra vigilant when ordering from abroad because UK law doesn't apply.

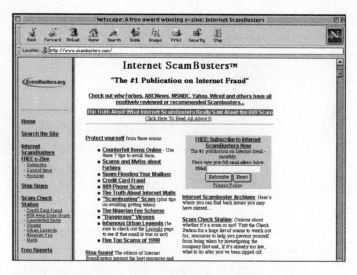

Protect yourself by keeping abreast of the latest scams at a site such as ScamBusters.

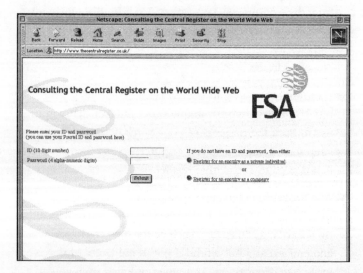

Is a financial services company properly authorised? Check on The Central Register.

⑨ If you're dealing with a financial product provider, check that it is fully authorized for its range of activities. You can check the company against the **Financial Services Authority**'s central register (**www.thecentralregister.co.uk**).

⑩ If a website is offering something that looks too good to be true, it probably is. Treat it with extreme caution. For news on the latest net frauds and scams try sites such as **Internet Fraud Watch** (**www.fraud.org**) and **Internet Scambusters** (**www.scambusters.com**).

⑪ Use a credit card to pay online. The card issuer is obliged to refund you under Section 75 of the Consumer Credit Act if the goods fail to arrive or are damaged.

Who's in charge of the net?

Well, no-one really. For the purists out there, that's its beauty. It embodies the principles of free speech in their purest form. For others it appears to be a chaotic nightmare, a haven for the depraved and the psychopathic. The techies who charted the net's rise from humble, innocent beginnings in the seventies, see the rise of rampant commercialism on the web as a sad diminution of its ethos of free exchange. In the US, home to the world's most dedicated conspiracy theorists, the net is often portrayed as the last truly free place on earth, in an age increasingly dominated by governmental and commercial surveillance of all aspects of our lives.

In many ways it is a strong force for democracy and freedom of expression. Tibetan monks, persecuted by Chinese occupying forces, have often used the net to communicate with the rest of the world, as have other downtrodden minorities. As a breathtakingly powerful tool for mass communication, the net can empower people by giving them information that they may not have been able to get hold of before. More than any other technological development it has shown that information is power.

So the net is still largely self-regulating, with businesses and service providers working out amongst themselves how best to deal with problems of security and authenticity. Governmental interference is not welcomed by the net industry, the need for businesses to find common security standards to facilitate e-commerce has led to an inevitable increase in the part that governments have to play. The web is transforming economies now, not just helping academics to swap ideas.

There is a constant battle between the libertarians and those who would regulate the net far more tightly, citing the ubiquity of pornography and other obscene or anti-social content to bolster their case. The problem is that as the net expands to become a global phenomenon, reaching Communist countries as well, Western ideas concerning freedom of speech don't go down too well.

Generally governments have been happy to sit and watch developments, seeing how far existing legislation can be stretched to cover online activities. For the most part, the web is just another publishing medium. So existing laws to do with libel, for example, apply equally online as offline. But there are some unique problems, many of them to do with copyright. The web has made it very easy to copy and distribute images and sound in clear breach of copyright law. Yet the net industry still believes technology will find solutions to these problems without the need for governments to impose cumbersome laws and potentially stifle business.

It is in the area of security that governments have made their presence felt. Law enforcement agencies generally want to retain the ability to open encrypted mail and get behind security firewalls when necessary in the name of national security. Civil libertarians argue that such powers would be open to too much abuse. Other commentators believe that try as it might, the government simply won't be able to control encryption because the technology is already out there and well established. Although the export of 128-bit encryption from the US is restricted to certain transactions (such as online banking) it is widely available, as is even stronger encryption, if you know where to look. The cat was let out of the bag a long time ago.

Chapter 9

Creating Your Own Web Pages

Introduction

M̲ore and more people are taking the opportunity to
jump on to the net and make their own contribution to
this global community by creating their own web page. If you
want to share your wisdom with the rest of the surfing world,
advertise your business, or even sell things, then your own
website is a must.

Many ISPs now offer free web space to their customers,
plus advice on how to use it. There are also plenty
of websites dedicated to helping you learn how to set
yourself up. For example, **Freewebspace.net**
(**www.freewebspace.net**) reviews over 400 free web space
providers.

AOL, the world's largest ISP, claims you can get your own
web page up and running in 20 minutes flat. Just don't
expect design awards to follow. They make it easy for you by
using a pre-formatted template. You can include your
personal details, enter whatever text you like, add links to
other pages and even add sounds. If you have a picture
scanner you can even upload a picture of yourself.

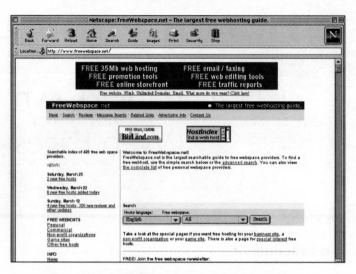

Reviewing the whole world of free web space (www.freewebspace.net).

Yahoo!'s **Geocities** (http://geocities.yahoo.com/home/)
offers free Web space to members which you can place in
any of 41 themed communities. You are also given the tools
to create your page. In return, Geocities asks you to include a
banner advert for their service on your site and a link back to
their home page. Similar free services can be found from
FortuneCity (www.fortunecity.com) and Lycos's **Tripod
(www.tripod.lycos.com)**.

Learning HTML

But if you want to get serious about publishing your
own website you need to learn a little about HTML –
Hypertext Mark-up Language. All web pages are written
in HTML. You can create HTML documents using any
text editor, such as Notepad if you have a Windows-
based PC, or SimpleText if you have a Mac. The latest
versions of web browsers contain basic **HTML editors** to
help you get started.

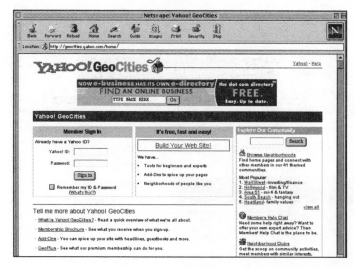

Online communities are a big growth area on the web.

Learning a bit about the fundamentals of HTML is quite a good idea. To denote the various elements in an HTML document, you use **tags**, which are instructions to your computer on how to format the text and images. These comprise a left angle bracket <, a tag name, then a right angle bracket >. For example, <H1> and </H1> would start and end the tag instruction for your main header.

Every HTML document should contain standard HTML tags, such as headings, body text, paragraphs, lists and other elements. First you have to tell the computer that this is an HTML document by writing <html> at the start. Other required elements are <head>, <title>, and <body> tags and their corresponding end tags. To centre a paragraph in your page you would write <P ALIGN=CENTER> at the start of the text and then </P> at the end. You can obviously format your elements any way you like with the relevant tags.

As with learning any new language, it can be daunting at first. But it's just a question of learning the vocabulary. One of the best ways to learn HTML is to see how other people have used it. You can do this by opening up your browser, going to a website you particularly like, and choosing a relatively simple page. In your browser click on 'View', then 'Source' or 'Page Source'. You then see a window with all the HTML commands used to create that page. You may be surprised how few commands it can take to create a relatively complicated-looking page. Then again, if you choose a complicated page – your ISP's home page, for example – the HTML commands can run to several pages if you print them off.

There's no point going into all the ins and outs of HTML in a general guide like this. If you really want to get to grips with the language, you can look at sites on the net that explain it in detail.

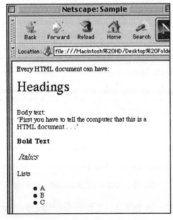

The HTML coding (right) for a simple web page (left).

What you see is what you get (WYSIWYG)

Website design starts to get complicated when you want to introduce dynamic features such as animated colour graphics, sound, or video. Luckily there are web authoring tools that enable you to create pages visually rather than writing large swathes of HTML code. These are called WYSIWYG editors – What You See Is What You Get – such as Microsoft FrontPage or Macromedia's Dreamweaver. You can usually find evaluation versions of these packages on free CD-ROMS accompanying computer and net magazines.

Bear in mind that these packages are designed to meet the needs of professional designers, so you'll need to spend a long time getting to know how they work to get the most out of them. If you simply want a web page as an alternative business card, you don't need such sophisticated tools.

Also, the more you stuff your site with complex graphics and groovy sound files the more web space you use up. Before too long you may find that you've used up the web space allocated to you by your ISP. The more sophisticated you become with your web page designs the more likely it is that you will encounter newer versions of HTML, such as DHTML and XML, and other programming and scripting languages such as Java, JavaScript, ActiveX, and CGI. This is seriously complicated stuff and is unlikely to affect the new user, so we don't have to go into it here, thank goodness.

Design tips

Generally speaking, the simpler you keep your web pages, the easier they will be to understand and the faster they will download into people's browsers. And let's face it, unless you've got something pretty compelling to offer, you're going to need every enticement to make people access your site. Offering a simple, no-fuss site that is quick to load certainly helps.

If you're going to have a website you may as well do it properly. That doesn't necessarily mean opting for all the latest fancy graphics and scripting tools. It means keeping it up to date and trying to fill the site with information that is useful, entertaining or both. It can be very annoying for surfers when they spend time perusing a site only to discover eventually that the information hasn't been updated since 1996 and is virtually useless. So put the date of the last time you updated your page in a prominent position.

T I P

The main rule to follow when designing a website for the first time is KEEP IT SIMPLE!

Earning trust

One of the problems the net faces is authenticating and verifying the information and opinions scattered across all these millions of websites. When surfers land on a site purporting to be the world's leading resource on Persian cats, say, how do they know that the author isn't an ignoramus posing as an expert, or worse, just dangerously deranged? The truth is that we don't. So if you're imparting information or opinions to the world, readers of your site are likely to be somewhat sceptical. And who can blame them?

If you want people to trust you and believe you, make an effort to establish your credentials. Tell people about yourself, maybe give a short career résumé, and give a contact address, telephone number, or e-mail. You've got to persuade strangers to your site that you know what you're talking about. Obviously, if you're worried about protecting your privacy, you should be selective in how much personal information you give away, and maybe stick to providing your e-mail address only.

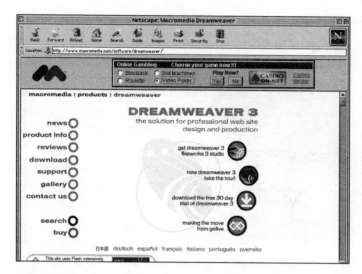

A number of products exist that allow you to create a professional-looking website.

State your sources if you are using research to back up an argument and clarify whether your views are mere personal opinion. Also bear in mind that although the web is often compared to the Wild West without the bullets, you are still governed by the same libel laws as other published material. If you use your web page as a platform to accuse your enemies of unspeakable practices you may well end up before the beak.

Web authoring tools

Macromedia's DreamWeaver

> **www.dreamweaver.com**

Microsoft's FrontPage   **www.microsoft.com/ frontpage**

NetObjects' Fusion   **www.netobjects.com**

HomeSite   **www.allaire.com**

Adobe GoLive   **www.adobe.com**

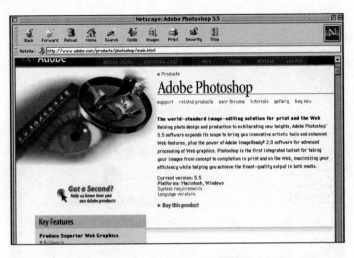

If you want to include complex graphics in your website you may need to use a separate graphics package.

Web page design advice

Web Developer	**www.webdeveloper.com**
eFuse	**www.efuse.com**
WebMonkey	**www.webmonkey.com**
Builder.com	**www.builder.com**
Developer.com	**www.developer.com**
Link Exchange	**www.linkexchange.com**
Web Site Garage	**www.websitegarage.com**
Davesite	**http://davesite.com/ webstation/html**
University of Leicester	**www.mcs.le.ac.uk/html/ BasicHTML.html**

Editing web graphics

Adobe PhotoShop	**www.adobe.com/photoshop**
Macromedia Fireworks	**www.macromedia.com/ software/fireworks**
Hemera Technologies NetGraphics Studio	**www.hemera.com**

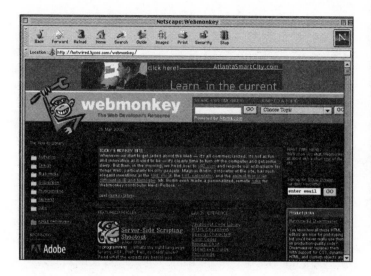

If you want to get serious about website design there are a number of excellent online resources, such as WebMonkey (above) and Web Site Garage (below), at your disposal.

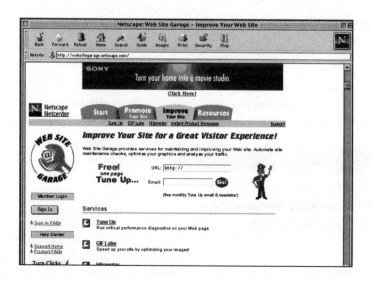

Where should I keep my website?

When you've designed your website, you don't have to keep it on your ISP's server. There are plenty of **web hosting services** around. These are companies that look after your website for you, store your web pages, and ensure that your site is available to the public 24 hours a day, seven days a week. Should you want to turn your website into a commercial venture, a web hosting service can usually handle online payments, too.

If you want your website to have a distinctive name, it's essential that you register the domain name. That's the website name plus the abbreviation after the dot. You can do this via a number of domain name registration companies, such as **DomainsNet** (**www.domainsnet.com**), **Nicnames** (**www.nicnames.co.uk**), or **Websitez** (**www.websitez.com**). It usually costs between £40 and £100, depending on the level of other services they offer. It will often take a long time entering your proposed domain names until you find one that hasn't been taken. Most of the obvious ones were snapped up a long time ago. This is perhaps why so many web services have silly names.

Owning your own domain name gives you more freedom over your e-mail address, too. You can have 'anyname@your-domain-name.co.uk' and the company will forward e-mails sent to this address to your existing e-mail address.

A web hosting service will usually handle the domain name registration part for you, as well as e-mail and URL forwarding. URL forwarding means that surfers typing in your web address, based on your new domain

> **TIP**
>
> *Most ISPs will offer you at least 20MB of space, which should be plenty for a small-scale website.*

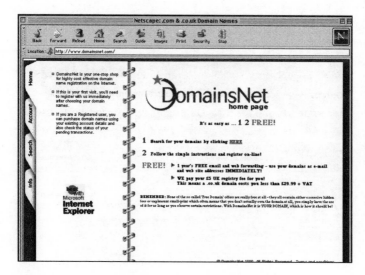

Register your domain name with a service such as DomainsNet (www.domainsnet.com).

name, will first go to the hosting company's servers before being bounced on to wherever you've decided to keep your website.

Most companies will offer you at least 20MB of space – enough for a small-scale website. But if you're planning something major, you may need to have an entire computer dedicated to handling your site. This can cost up to £10,000 a year, so make sure yours is a serious commercial venture before you opt for this!

It is tempting to stick with the free web space offered by your ISP, and in most cases this will be good enough for people just wanting to have a bit of fun. But if you are more serious about establishing a web presence, reliance on your ISP can backfire in the long-term. For example, **Freeserve** (**www.freeserve.net**) doesn't allow you to use your own domain name on its servers – you have to accept whatever name it gives you. If you're set on having your own name,

If you are creating a website for fun, there are a number of services that will host your site for free.

you have to move your site somewhere else, which can be irksome to say the least.

Another point to bear in mind is that not all ISPs will allow you to conduct commercial activities on websites that they host. If you plan to move into business, check first with your ISP and any proposed hosting service that they do allow this. Sometimes ISPs will not allow you to use the software necessary to run your own commercial website – product catalogues, for example, or application forms, for fear that you'll crash their servers and put all their members' noses out of joint.

TIP

Not all ISPs will let you conduct commercial activities on websites that they host. If you plan to move into business, check first with your ISP and any proposed hosting service that they do allow this.

A free service such as **Freenetname** (**www.freenetname.co.uk**) may also seem tempting at first. But you have to dial into its service to pick up your e-mail or edit your site. You can't access it from another ISP. And if you want to move your site elsewhere, Freenetname charges you nearly £100 for the privilege.

As with most things in life, you generally get what you pay for. The more you cough up, the more comprehensive will be the list of services on offer, the web space provided, and the freedom to run things the way you want. For around £200 a year you can get your own domain name, plus shared space on a web hosting company's server. But there may be limitations on the amount of **traffic** your site can receive. Traffic is the term used for the flow of visits websites receive from browsing surfers.

It is possible to host your site in the US. There is greater competition over there so prices are generally lower. But your

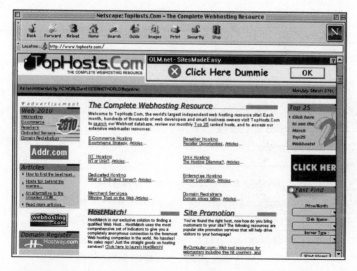

Before you decide where to host your website, go to a site such as www.tophosts.com for reviews of the various web hosting services.

pages may download more slowly because of the greater distance they have to cover.

If you're shopping around for a web hosting service, go to **www.findahost.com** and **www.tophosts.com** for reviews and comparisons. Some internet magazines also publish the result of monthly tests on web hosting companies.

Here's a checklist of what the perfect web hosting service should offer:

- technical support 24 hours a day, seven days a week

- website design advice

- a chance to visit their premises and check the security of the site

- back-ups in case of server corruption or theft

- business references and favourable reviews

- online shopping service

- fast connection times

- the facility to handle more sophisticated pages that may include video and audio

- no restrictions on the amount of traffic your site can receive

- low charges if you decide to move your site to another host in future

How will people know about my website?

It is surprisingly common for companies who should know better to set up websites and then expect a flood of visitors as soon as they're up and running. If only it were that simple. Cyberspace is a very big place and the chances of anyone landing on your site at random are miniscule. Don't forget, there are over a billion web pages out there!

One way to publicize your website is to get other websites to publish a link to yours. You can usually do this for free if you come to a reciprocal arrangement – I'll link to yours if you link to mine. A fancy link including graphics is called a **banner ad** and schemes such as **Link Exchange** (**www.linkexchange.com**) arrange for smaller websites to include each other's banner ads on their pages. It's also a way to make money if you include links to commercial ventures. This is called affiliate marketing. If someone comes to your site and clicks on one of these commercial links on your pages – known as a **click though** – you get a small percentage of any sale that results. **SimpleSite** (**www.simplesite.co.uk**) has a list of all the affiliate schemes you could sign up to.

Big companies pay millions of pounds to advertise their sites on popular portal sites and search engines because they know millions of people are going to visit those sites and see their ads. It's the same as conventional newspaper, magazine and television advertising. The more traffic you can direct to your site, the more attractive it will appear to potential advertisers and the more money you'll make.

Presuming you're not in the big league yet – and if you are you shouldn't really need to read this guide! – another common way to advertise your site is to submit details of your URL to all the search engines and directories. It's free. The problem is that they all have different ways of indexing sites. Some will just pick out key words on your site, others will index it in full, others will accept summaries of what your site contains. You should spend some time investigating what methods they use – **Search Engine Watch** (**www.searchenginewatch.com**) is a useful site to look at.

If you haven't the time or the inclination to do all this yourself, you can leave it to other companies to submit your website and/or promote it for you (*see list on page 201*). If

By including links to commercial companies you might be able to make money through your website.

you're lucky, your site may be picked up by the more thorough search engines anyway, even if you do nothing. Their 'spiders' or 'bots' are constantly scouring the entire web for new pages to add to their databases.

Getting your site to come up when people search on related words is a black art – it mystifies lots of companies. Mostly it involves using meta tags – key words that help to identify the main subjects covered by your site. These are the words that search engines will focus on when trying to sort out searches for relevance.

TIP

When you set up a website, don't forget to include the address on all your traditional paper advertising – letterheads, business cards, flyers, invitations and so on.

Pornography sites in particular have shamefully picked up on this ruse and often include innocuous words among their meta tags to try to lure unsuspecting people into their lairs (*see* **Safe Surfing**, *page 159* for more on how to screen out adult content on the web). You often hear at dinner parties how a quite innocent search threw up some far-from-innocent material.

You could also try advertising your website in newsgroups and mailing lists (*see* **Making the Most of the Web**, *page 129*). Obviously it makes sense to target your ads at groups with similar interests. Then there's the trusty paper-based advertising methods of old, such as letter-heads, business cards, printed T-shirts and other merchandise. Many 'dot com' companies still believe such offline advertising to be the most effective.

Some well-known web hosting services

Corpex	www.corpex.com
WebFusion	www.webfusion.co.uk
Magic Moments	www.magic-moments.com
Global Internet	www.global.net.uk
EasySpace	www.easyspace.co.uk
Rapidsite	www.rapidsite.co.uk

Submitting your website

Addme!	www.addme.com
Net Submitter	www.sutzon.com/ SubmitSpider
Broadcaster	www.broadcaster.co.uk
Submit It	www.submit-it.com
Exploit	www.exploit.com

Promotion and ranking services

Web Promote	www.webpromote.com
Web Position Agent	www.webposition.com
RankThis!	http://rankthis.webpromote.com

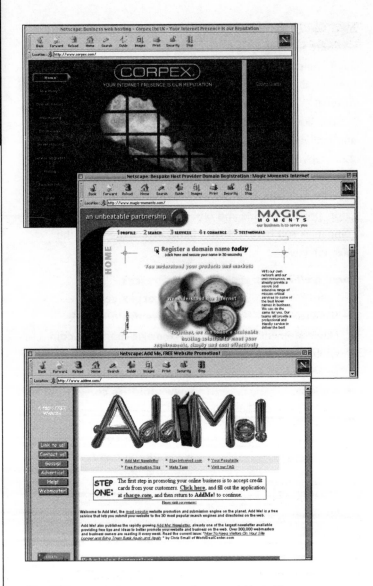

*Top and centre: Web hosting services Corpex and Magic Moments.
Bottom: Go to sites such as AddMe! to subit your own website.*

A Brief History of the Internet

The internet is an unstructured worldwide network of computers and computer networks linking millions of users, including private individuals, businesses, academic institutions and governments. It facilitates the exchange of information, commerce, entertainment, and financial transactions.

The main services on the Internet are e-mail, the World Wide Web (the web), Usenet, and FTP (File Transfer Protocol). The computer networks are linked by telephone lines, fibre optic cables, microwave relays and satellite. Networks and private users mostly gain access via an Internet Service Provider (ISP), which they dial up from their personal computers using a modem.

Computers talk to each other in a number of standard languages or protocols, but the backbone is Transport Control Protocol/Internet Protocol (TCP/IP). Other common languages and protocols are HyperText Markup Language (HTML), which is the language of the World Wide Web, and HyperText Transfer Protocol (HTTP) by which files are transferred across the web.

One of the most surprising things about the internet is how young it still is and yet how rapidly its use has grown. The term 'internet' did not even become common currency until 1982. In the six years from 1984 to 1989 the number of internet hosts grew a hundredfold to 100,000, rising to one million by 1992, and to over 10 million in 1996. There are now an estimated 100 million internet users worldwide.

The internet was dreamt up by the US military as a way of protecting valuable information on computers from nuclear attack. Instead of relying on a single two-way link, the idea was that the computer would have many other links available should one link be destroyed.

A system called 'packet switching' was developed whereby the originating computer divides a message to be sent into small parcels, then bundles each parcel into small electronic envelopes or packets containing the address of the destination computer. The packets are then sent over the internet by the quickest available route.

This means that packets making up a message may reach the destination via different routes, depending, for example, on the degree of traffic from other users. The first packet-switched network was designed in 1969 and installed in four US universities in 1970.

This first network was called ARPAnet, after the Advanced Research Projects Agency of the US Department of Defense which had devised and funded the system. Before long, protocols were agreed, whereby computers would transfer data files in a standard format. The main protocols, TCP/IP, were introduced in 1974. E-mail followed in 1976.

The seventies also saw the development of personal computers (PCs) and the rise of other computer networks, such as the Computer Science Research Network (CSnet), which eventually found a way of linking up with each other in 1982 to form the first version of the internet.

Soon it was not just academics communicating with each other electronically, but businesses and individuals, as PC ownership exploded throughout the eighties. A catalyst for the surge in the popularity of the internet in the nineties was the relaxation of US National Science Foundation's rules governing how its important NSFnet network could be used. Commercial and personal use was allowed for the first time in 1992.

The World Wide Web, a particular service on the internet, was made possible by Tim Berners-Lee, a Briton working at CERN, the nuclear research facility in Switzerland. He wanted a way to link up academic physicists and developed the hypertext language system. Mosaic was the first popular web browser to be developed, soon followed by Netscape's Navigator, and this made the web accessible to millions more people.

Now the internet can carry images, video and sound, as well as text. But images contain far more data than text, and so take up more room in the pipeline, whatever medium that may be. As more and more information needs to be carried over connections that can only take so much capacity, the speed with which people can access information is slowing down, especially when millions of people use the internet at the same time.

So the next major challenge is to increase bandwidth – the speed at which information can be sent down the wires – and the efficiency with which information is sent over the network. A dedicated fibre optic cable, for example, can have a bandwidth 1,000 times greater than standard copper twisted-pair telephone wires. Websites are becoming increasingly interactive, with the introduction of online transaction capabilities. New technologies that can customize web content to suit users' tastes are also making the internet a more manageable place.

Glossary

ADSL – Asynchronous Digital Subscriber Line – a way of
sending digital data over the conventional telephone network
at speeds up to 40 times faster than the fastest modem.
Downloading data is faster than uploading.

aggregator – website that brings together in one place lots of
links to other websites. It may offer its own entertainment,
information and shopping services, too.

applet – a small helper program (usually written in Java
programming language) that is downloaded from the net into
your Web browser as and when necessary.

backbone – high capacity data motorways that link other
networks together.

bandwidth – the speed with which data can be sent across
the network. The bigger the bandwidth the more data can be
sent at one time and so the faster the system.

banner ad – graphical link on a web page usually advertising
a commercial service.

BBS – Bulletin Board System. A system that allows you to post and receive messages and download files.

beta programs – usually free pilot software programs released to the public by software companies for testing and evaluation purposes.

bits – the smallest pieces of information a computer deals with.

bits per second (bps) – the standard measure of speed in transmitting data. A modem with a speed of 56 kilobits per second can download data at up to 56,000 bits per second, though it rarely achieves such maximum performance.

byte – a piece of digital information containing eight bits.

Boolean operators – words such as AND, NOT, OR which help you be more specific when using a search engine.

broadband services – high-speed connections to the internet, 10 to 100 times faster than the fastest modem.

browser – software program, such as Netscape Navigator or Internet Explorer that helps you surf the World Wide Web.

cable modem – a gizmo that connects your computer to a fibre optic cable as opposed to the usual twisted copper wires of the conventional telephone network.

CD-ROM – compact disk read-only memory. A disk that can store large amounts of information.

channels – another word for web services usually incorporated into web browsers, portal sites and search engines.

chat room – forum for sending and receiving instantaneous messages to any number of other people who are joining the 'conversation'.

click through – the process of clicking on a link or advert on a web page and going to another site.

client – another name for a helper software program, such as a web browser or FTP manager.

compressed file – a file that has been crunched down to take up less memory so that you can download it quicker. To read these files you need to decompress them using a program such as WinZip.

cookie – a small file downloaded on to your computer that enables a web company to track where you've been on the net.

cyberspace – abstract description of the world of the net and the web.

digital signature – a way of authenticating the origin and authorship of digital documents.

directory – a categorised list of links to websites.

Domain Name System (DNS) – the way of translating domain names, such as 'Excite.com', into Internet Protocol addresses that identify individual computers by a unique series of numbers. These domain names are managed and distributed by 'nameservers'.

download – transfer files and web pages on to your computer.

dynamic IP – an internet protocol address for your computer that changes every time you go online, helping to prevent websites identifying you as a repeat visitor. Not all ISPs support dynamic IP.

e-mail – electronic mail. Messages that can be sent from one computer to another using a unique address. You can also attach files to your messages.

encryption – a way of jumbling up digital data so that it cannot be read by anyone who does not have the key to unscramble the message.

File Transfer Protocol (FTP) – standard for transferring files across the net.

FAQ – frequently asked questions. A document on a website or newsgroup that answers common queries.

firewall – a security system that prevents outsiders gaining access to a particular network of computers.

fix or **patch** – extra code added to a software programs to improve it or iron out a glitch

flaming – sending abusive messages via e-mail or a newsgroup.

forwarding service – e-mail server that will redirect mail sent to one address to another address of your choice.

freeware – software that you can download for free.

Gigabyte (Gbyte) – unit of measurement for computer memory, roughly one billion bytes.

hit – a file found as a result of search request.

home page – normally the first page you see when accessing a website.

host – any computer open to external online access, usually providing services for other computers. Also known as a server.

HTML – HyperText Mark-up Language. The standard language for displaying text in web pages.

HTML editor – software that helps you translate standard word processing text into HTML code for web pages.

HTTP – HyperText Transfer Protocol. The standard for transferring HTML documents across the net. Web addresses always begin with the letters http.

HyperText – text or graphics that when clicked on take you to another site called a hyperlink.

Internet Protocol (IP) address – a unique address assigned to every computer on the net that usually consists of four sets of numbers separated by dots.

ISDN – Integrated Services Digital Network. A way of transferring digital data over standard telephone wires that gets rid of the need for modems. Can achieve speeds of up to 128Kbps.

ISP – Internet Service Provider.

leased line – permanent and dedicated high-speed connection to the internet (2Mbps usually), mostly adopted by businesses.

mailer – another word for e-mail software.

mailing list – service that delivers information or news to subscribers via e-mail.

Messenger – Netscape's e-mail program that comes with the Navigator web browser.

Megabyte (MB) – a million bytes.

meta-search engine – search engine that trawls several other search engines as well as its own database for a more comprehensive search.

modem – gizmo for translating digital data into analogue format for transfer over the telephone lines.

multimedia – the integration of sound, text, pictures and video in digital communication.

Navigator – another word for browser, but usually shorthand for Netscape Navigator, the second most popular browser.

net – short for internet.

netiquette – code of behaviour for net surfers whilst online.

newsgroup – a particular discussion group where people can read and post messages on a specific topic. The biggest collection of such newsgroups is Usenet.

newsreader – browser for reading and organizing Usenet messages.

notifier – web service that will e-mail you when a particular web page you're interested in is updated or changed.

offline – not connected to the net.

online – connected to the net.

Outlook Express – a Microsoft e-mail program that comes with its Internet Explorer web browser.

PDF files – files created using Adobe Acrobat that require Acrobat Reader to view them.

plug-in – an additional program downloaded on to your web browser that allows you to watch a video, for example, or listen to live radio.

portal – a general-purpose website that may contain its own entertainment and information services as well as links to many others. It usually incorporates a search engine or directory.

ripper – a program that transfers music from a CD onto your hard drive.

search engine – a program that interrogates the net for files or web pages containing or relating to words or phrases you enter into the search box.

Secure Sockets Layer (SSL) – system developed by Netscape for encrypting online payment details.

server – *see* host

shareware – software that is usually free for an evaluation period, after which you have to pay a registration fee.

signature file – a file usually containing your contact details that you can include automatically with your e-mails.

SMTP – Simple Mail Transfer Protocol. The standard for sending e-mail across the net.

spam – unsolicited e-mail.

streaming audio and video – listening to music and radio, or watching video, while the data is in the process of being downloaded, rather than having to wait for entire files to be downloaded first.

surfing – exploring files on the net.

tags – instructions on how to format text and images in HTML for web pages.

TCP/IP – Transport Control Protocol/Internet Protocol – the standard way computers talk to each other on the internet.

thread – a discussion within a Usenet discussion group.

traffic – the term used for the flow of visits sites receive from browsing surfers.

upload – transfer files, documents and web pages from your computer on to the net.

URL – Uniform Resource Locator. The standard way of identifying any service on the net, such as
http://www.sunday-times.co.uk

Usenet – the collective name for discussion newsgroups on the internet.

vCard – Netscape Messenger's name for a signature file.

virus – a computer program designed to corrupt information on another computer.

web hosting service – company that will store and maintain websites on their own servers, register domain names, and offer other services, such as e-mail and URL forwarding.

web server – computer that stores web pages that can be accessed by other computers.

Wireless Application Protocol (WAP) – standard for making web pages suitable for viewing on hand-held devices, such as mobile phones, pagers and personal organizers.

World Wide Web (the web, or www) – collective name for all documents on the Net written in HTML that can be read using a web browser.

zip – to compress data so that it takes up less space and can be downloaded faster. Once received, zipped files need to be unzipped by a software program or else they can't be read.

Index